59 Springer Series in Solid-State Sciences

Edited by Peter Fulde

Springer Series in Solid-State Sciences

Editors: M. Cardona P. Fulde H.-J. Queisser

40 **Semiconductor Physics –**
An Introduction By K. Seeger

41 **The LMTO Method**
Muffin-Tin Orbitals and Electronic Structure
By H. L. Skriver

42 **Crystal Optics with Spatial Dispersion, and Excitons**
By V. M. Agranovich and V. L. Ginzburg

43 **Resonant Nonlinear Interactions of Light with Matter**
By V. S. Butylkin, A. E. Kaplan, Yu. G. Khronopulo, and E. I. Yakubovich

44 **Elastic Media with Microstructure II**
Three-Dimensional Models
By I. A. Kunin

45 **Electronic Properties of Doped Semiconductors**
By B. I. Shklovskii and A. L. Efros

46 **Topological Disorder in Condensed Matter**
Editors: F. Yonezawa and T. Ninomiya

47 **Statics and Dynamics of Nonlinear Systems**
Editors: G. Benedek, H. Bilz, and R. Zeyher

48 **Magnetic Phase Transitions**
Editors: M. Ausloos and R. J. Elliott

49 **Organic Molecular Aggregates,** Electronic Excitation and Interaction Processes
Editors: P. Reineker, H. Haken, and H. C. Wolf

50 **Multiple Diffraction of X-Rays in Crystals**
By Shih-Lin Chang

51 **Phonon Scattering in Condensed Matter**
Editors: W. Eisenmenger, K. Laßmann, and S. Döttinger

52 **Superconductivity in Magnetic and Exotic Materials**
Editors: T. Matsubara and A. Kotani

53 **Two-Dimensional Systems, Heterostructures, and Superlattices**
Editors: G. Bauer, F. Kuchar, and H. Heinrich

54 **Magnetic Excitations and Fluctuations**
Editors: S. Lovesey, U. Balucani, F. Borsa, and V. Tognetti

55 **The Theory of Magnetism II**
Thermodynamics and Statistical Mechanics
By D. C. Mattis

56 **Spin Fluctuations in Itinerant Electron Magnetism** By T. Moriya

57 **Polycrystalline Semiconductors,**
Physical Properties and Applications
Editor: G. Harbeke

58 **The Recursion Method and Its Applications**
Editors: D. Pettifor and D. Weaire

59 **Dynamical Processes and Ordering on Solid Surfaces**
Editors: A. Yoshimori and M. Tsukada

Volume 1–39 are listed on the back inside cover

Dynamical Processes and Ordering on Solid Surfaces

Proceedings of the
Seventh Taniguchi Symposium,
Kashikojima, Japan, September 10–14, 1984

Editors: A. Yoshimori and M. Tsukada

With 89 Figures

Springer-Verlag
Berlin Heidelberg New York Tokyo 1985

Professor Dr. Akio Yoshimori

Faculty of Engineering Science, Osaka University, Toyonaka,
Osaka 560, Japan

Professor Dr. Masaru Tsukada

Department of Physics, Faculty of Science, University of Tokyo,
Hongo 7-3-1, Bunkyo-ku, Tokyo 113, Japan

Series Editors:

Professor Dr. Manuel Cardona
Professor Dr. Peter Fulde
Professor Dr. Hans-Joachim Queisser

Max-Planck-Institut für Festkörperforschung, Heisenbergstrasse 1
D-7000 Stuttgart 80, Fed. Rep. of Germany

ISBN-13:978-3-642-82425-8 e-ISBN-13:978-3-642-82423-4
DOI: 10.1007/978-3-642-82423-4

Softcover reprint of the hardcover 1st edition 1985

2153/3130-543210

Preface

This volume is the proceedings of the Seventh Taniguchi International Symposium on the Theory of Condensed Matter. The symposium was held for five days from September 10 to 14, 1984 at Kashikojima, Mie, Japan. Dynamical processes and ordering on solid surfaces are the subjects of the symposium. About twenty participants stayed together at Shima Kanko Hotel, the symposium site, during the period. The intense and productive discussion in the bright seaside atmosphere of Kashikojima is believed to have been impressive to all the participants.

Dynamical processes on solid surfaces are the target of recent theoretical efforts in surface physics. Even if some of them are still in their infant stage, important aspects begin to appear and vital concepts start to shape themselves. Some topics in the symposium were the energy transfer related with internal degrees of freedom of molecules, attempts to go beyond the trajectory approximation, charge transfer and energy transfer between particles and solid surfaces, and related fundamental problems like adiabatic potentials and electronic structures. In particular, really actively discussed was the time-dependent Newns-Anderson model without and with the intraatomic Coulomb interaction and sometimes with the interaction to the surface plasmons or phonons. Surface effects on the optical processes were discussed with great interest, such as the ABC-related problems of exciton polaritons and rare gas adsorbates on metal surfaces. The ordering problem, such as reconstructions of adsorbate superlattices, has received a great impact by the discovery of the scanning tunneling microscope technique. This is seen in the recent development of the investigation of the Si(111)7x7 reconstructed surface. Topics discussed were the ground state in the lattice gas model, rare gas monolayers on graphite, and the W(100) surface reconstruction.

As the prefaces in the previous proceedings of the Taniguchi International Symposia have mentioned repeatedly, Mr. Toyosaburo Taniguchi's support to natural and human sciences should be gratefully acknowledged, by which this symposium was made possible. Experiences and stimulations, which younger participants have had, are immeasurably important not only for themselves but hopefully for this particular field of physics. We would like to mention our deep thanks to Mr. T. Taniguchi and the Taniguchi Foundation.

December, 1984 *A. Yoshimori and M. Tsukada*

Foreword
Speech by Toyosaburo Taniguchi

Dr. Kubo, Distinguished Guests, and Friends:

I am very happy, pleased, and honored to be here this evening with so many distinguished guests, friends, and scholars from different parts of the world.

The Taniguchi Foundation wishes to extend a warm and sincere welcome to the many participants of the Seventh International Symposium on the Theory of Condensed Matter; this was inaugurated six years ago through the dedicated efforts of Dr. Ryogo Kubo, who is gracing us today with his presence.

We are deeply indebted to Dr. Kubo, Dr. Yoshimori, and their associates for their unsung and unstinting efforts which have been so crucial in rendering this particular symposium possible.

We are convinced that what makes our symposium a success is neither the Foundation nor its founder, but the outstanding quality and fine personality of the participants.

In this sense, your words of praise about the symposium, if any, should be directed to them and to those who have directly or indirectly lent their support to this program.

I am willing to accept your thanks if your words of praise are given me for the dishes prepared by this top-level hotel in Kyoto, for I have chosen them personally.

To date, we are proud to note that there has been a total of 44 participants in this division from 9 countries:

> Canada, Denmark, France, Ireland, Israel,
> Switzerland, U.K., U.S.A., West Germany,
> with 100 participants from Japan.

And, we are all friends.

Whenever I meet with the participants in our symposia, both young and old, I am deeply impressed by the unselfish and sincere dedication and sense of mission they display, in pursuing the vocation they are in.

To those younger people who were made welcome by world-famous scholars as their friends, we offer our hopes that even after ascending the climb to academic fame within, say, the next 10 years from now, or 20 or 30, you will perpetuate your ties, join forces, and help to forge closer bonds of friendship and cooperation in a manner that will make a major contribution not only to academia but also to permanent world peace.

And, it is our hope that our symposia be continued, as long as the fund permits, say, for about several years, unless new and unexpected situations arise.

In conclusion, we sincerely trust that all the participants, both Japanese and foreign, will return home bearing heartwarming memories of the symposium and of the pleasant times we have been enjoying together.

Thank you

Participants in the Taniguchi International Symposium, September 10–14, 1984 (at Kashikojima, Japan)

Seated (*left to right*)

A. Yoshimori W. Brenig J. Villain K. Schönhammer N.D. Lang D.M. Newns J.K. Nørskov R. Kubo J. Kanamori M. Tsukada

Second row (*left to right*)

S. Sugano Mrs. A. Nørskov K. Shinjo H. Ueba T. Matsubara A. Okiji K. Makoshi E. Hanamura N. Shima T. Inaoka

Third row (*left to right*)

K. Cho C. Satoko H. Nagayoshi Y. Ishii R. Kawai S. Shindo

Contents

Part I Dynamical Processes

Part I

Dynamical Processes

Quantum Theory of Rotational and Vibrational Energy-Transfer at Surfaces

W. Brenig, H. Kasai[+)], and H. Müller

Physik Department, TU München, D-8046 Garching-München, Fed, Rep. of Germany

A new method for treating the coupled channel problem for inelastic molecule surface scattering is presented and applied to the problem of rotational,vibrational and combined rotational-vibrational energy-transfer. The results are compared with those of the so-called sudden approximation and simple ways of improving this approximation are discussed.Experimental rotational state distributions are presented in the context of theoretical results. Phenomena such as "rotational rainbows" and "desorptive rotational cooling" will be discussed.

1. Introduction

A microscopic understanding of kinetic processes at solid surfaces requires a detailed knowledge of the various energy transfer processes occurring in molecule-surface interactions. Recently there has been growing interest in the dynamics of rotational energy-transfer,in particular of rotationally inelastic molecule-surface scattering.

For light molecules such as H_2,D_2,HD at low energies,angular[1] and energy[2] distributions can be used to extract rotational transition probabilities. For heavier molecules,e.g. NO, at higher energies laser-induced fluorescence has been used to determine the rotational state distribution of surface scattered molecules [3,4,5,6].

Interesting phenomena at low energies are the processes of rotationally mediated selective adsorption and the corresponding resonances in the scattering cross-sections; at higher incident energies one finds maxima in the rotational state distributions which could be identified as socalled "rotational rainbows" [4,6] and plateaus in the effective temperature of scattered particles [5] indicating what may be called "rotational cooling".

Theoretically one has investigated simple models which in many cases give surprisingly good semiquantitative descriptions of the experimental results. Rotational energy-transfer,of course, requires an anisotropic potential. The simplest model consists of a flat (no corrugation), rigid (no phonons) surface with a symmetric (no odd spherical harmonics) anisotropic potential. Such a model is already capable of describing the occurrence of a high energy rotational rainbow.With appropriate parameters the experimental position of the rainbow can be reproduced [8]. The existence of a second rainbow at low energies requires the introduction of a rather strong asymmetry [9,12]. With appropriate asymmetry and well depth of the potential one can reproduce the experimental positions of the two rainbows. The intensities and widths of the rainbow peaks, however, could not be described quantitatively.

[+)] On leave of absence from: Department of Applied Physics, Osaka University Osaka , Japan.

Perhaps both corrugations and vibrations of the surface have to be taken into account to describe these results correctly. Corrugations have been considered already in connection with the selective adsorption experiments because the combined effect of rotational and diffractional energy-transfer is essential for an understanding of these data [7].

Because of the high incident energy (up to about 1eV) of the rainbow experiments, because of the large mass ratio $m(NO)/m(Ag) \cong 0.3$ and because of the high surface-temperatures one expects large vibrational energy transfers with the phonons of the substrate. They have been discussed classically [13a] and quantum mechanically [13b]. The quantum mechanical results of Schinke and Gerber are, so far, only concerned with the Debye-Waller attenuation of zero phonon rotational intensities. One finds an increasing Debye-Waller factor with increasing rotational energy transfer. This leads to a shift of the rainbow position to higher energies. These effects are, however, not very large and do not depend strongly on temperature.

The combined rotational-vibrational energy-transfer deserves further attention. First of all the overall Debye-Waller factor is small. Thus most of the intensity is not in the zero phonon line but in the phonon "side bands". One would expect a decreasing rotational energy-transfer with increasing vibrational energy-transfer: The phonon side bands redistribute rotational intensity to lower angular momenta. This is just the opposite of the Debye-Waller attenuation. Furthermore the results of [13b] were obtained using the so-called sudden approximation [10,11]. It is known from the discussion of the vibrational case alone, that this approximation overestimates the phonon effects considerably for larger mass ratios and energy transfers [14].

In this contribution we plan to present a new method for treating the coupled channel problem for both rotational and vibrational energy-transfer, to propose simple improvements of the sudden approximation, and using this to reconsider the rotational energy-transfer problem under the simultaneous influence of the substrate phonons.

2. Scattering theory in terms of Jost functions

This section will be mainly methodical in nature. It presents the method which we have been using to solve the coupled channel problem numerically. Rather than integrating the Schrödinger eq. directly starting at small distances from the surface, we use the integral equation method proposed by JOST and PAIS [15]. The iterative solution of this eq. turns out to be very simple for Morse potentials. It leads to a rapidly converging series for the Jost functions, from which the S-matrix can be obtained by a single matrix inversion.

We start out from the Schrödinger eq. in (ℓ,z)-representation, ℓ being the angular momentum quantum number and z the center of mass coordinate of the molecule perpendicular to the surface

$$\{-(\partial/\partial z)^2 - k_\ell^2 + v(z)\}\psi(\ell,z) = 0 \tag{1}$$

All energies are measured in units of $\hbar^2\alpha^2/2M$ where M is the mass of the molecule and α the range constant of the potential (reciprocal length unit). k_ℓ is the translational wave number given by

$$k_\ell^2 = k^2 - r\ell(\ell+1) \tag{2}$$

where k^2 is the total energy and r the rotational constant

$$r = M/I\alpha^2 \quad ; \quad I = \text{moment of inertia} \tag{3}$$

Negative k_ℓ^2 correspond to closed, positive to open channels. The two kinetic energy terms in (1) are diagonal in ℓ while the potential v(z) is a matrix:

$$v(z)\psi(\ell,z) = \Sigma\, v_{\ell\ell'}(z)\psi(\ell',z) \tag{4}$$

If we restrict ourselves to flat surfaces the azimuthal quantum number(m) is conserved.The matrix v(z) may depend on m but we do not explicitly indicate this dependence.The generalization to energy-transfers to additional degrees of freedom e.g. m, momenta parallel to the surface and vibrational quantum numbers is straightforwardly taken into account by additional labels of v(z). We are going to consider vibrational degrees of freedom later on.

Angular momenta pertaining to open channels only will be denoted by j. A state with incoming angular momentum j then will behave asymptotically as

$$\psi_j(\ell,z) \to (e^{-ik_j z}\delta_{\ell j}/\sqrt{k_j} - e^{ik_\ell z} S_{\ell j}/\sqrt{k_\ell}) \quad ; \quad z \to \infty \tag{5}$$

where $S_{\ell j}$ is the S-matrix. If one integrates each of the two terms on the r.h.s. of (5) down from large z towards the surface,one obtains

$$\psi_j(\ell,z) = f_j(\ell,z)/\sqrt{k_j} - \Sigma_{\ell'} f^*_{\ell'}(\ell,z)\, S_{\ell' j}/\sqrt{k_{\ell'}} \tag{6}$$

where the functions f(z) which we will call Jost functions obey the int.eqs.

$$f_{\ell'}(\ell,z) = e^{-ik_\ell z}\delta_{\ell\ell'} + \int_z^\infty \sin k_\ell(z'-z)v(z')f_{\ell'}(\ell,z')dz'/k_\ell \tag{7}$$

The S-matrix then is determined by the condition that the Schrödinger function (6) vanishes for z going to minus infinity

$$f_j(\ell,a) = \Sigma_{\ell'} f^*_{\ell'}(\ell,a)\, S_{\ell' j}\sqrt{k_j/k_{\ell'}} \quad ; \quad a \to -\infty \tag{8}$$

Practically one has to choose a sufficiently far below the classical turning point.We varied a until there was no change in S up to the tenth digit.

One advantage of the Jost functions as compared to the Schrödinger function is the fact that they are asymptotically independent of v(z). Since the Schrödinger eq. (1) is analytical in the strength of v the Jost function in contrast to the Schrödinger function and the S-matrix is analytical in the strength of v.The iterative solutions of the integral eqs.(7) thus converge for arbitrary potential strengths - in contrast to the Born series for the Schrödinger éq.-. In addition it was demonstrated in the original paper [15] by JOST and PAIS that,in practice, the iterative solution of (7) converges much more rapidly than the Born series for (1).

The iterative solution of (7) becomes very simple for Morse type potentials

$$v(z) = v\,\exp(-2z) - w\,\exp(-z) \tag{9}$$

where v and w are matrices independent of z. The z'-integration in (7) can then be carried out analytically for arbitrary order n of the iteration.The result is a power series expansion in exp(-z)

$$f_{\ell'}(\ell,z) = \sum_{n=0}^{\infty} f_{\ell'}(\ell,n)\, e^{-(n+ik_{\ell'})z} \tag{10}$$

and for the expansion coefficients f(n) one finds a two-step recurrence rel.

$$f_{\ell'}(\ell,n) = \frac{1}{n(n+2ik_{\ell'})+k_\ell^2-k_{\ell'}^2}\,(v\, f_{\ell'}(\ell,n-2)-w\, f_{\ell'}(\ell,n-1)) \tag{11}$$

The convergence of (10) is such that one roughly needs at least as many terms in the sum as there are nodes in the wave function.For a ten digits accuracy one needs four to five times as many terms.In practice,for incident energies of the order of 0.5 eV and well depths of the order of 0.2 eV we needed of the order of 300 terms. In order to avoid strong oscillations in the coefficients f(n) we found it useful to split off a factor exp(-aexp(-z)) from the Jost functions where $a^2 = \overline{v}$ is some average repulsive diagonal matrix element for v.For the remaining factor which we call f(z) again,one then obtains an expansion like (10).The expansion coefficients,however,now obey the recurrence relation

$$f_{\ell'}(\ell,n) = \frac{1}{N}\,\{(v-\overline{v})f_{\ell'}(\ell,n-2) - (w-a(2ik_{\ell'}+2n-1))f_{\ell'}(\ell,n-1)\} \tag{12}$$

where N stands for the denominator occurring already in (11).

3. Comparison of exact,approximate and experimental results

Figure 1 shows the rotational state distribution

$$P_{jj'} = |S_{jj'}|^2 \tag{13}$$

in particular for j=0 which were obtained using the formalism described in the section above,as compared to various approximations. The intensity appears to oscillate around two peaks corresponding to the two rotational "rainbows" [9,12]. These rainbows have a simple classical interpretation. Semiclassically the S-matrix is related to an angle-dependent phase shift

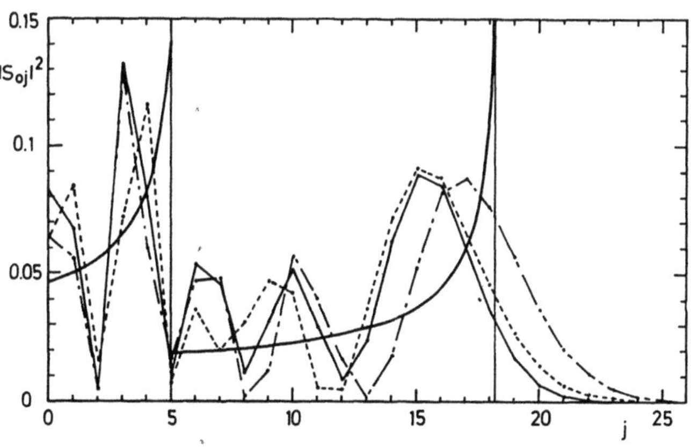

Fig.1 Rotational intensities for a potential $v=1+0.21P_1(\cos\theta)+0.14P_2(\cos\theta)$, w=0, α=1.4/Å E=0.5eV.———:exact and classical, — - —:quantum sudden appr., -----:"sudden"phase(17) multiplied by 0.92, initial j = 0.

$\eta(\theta)$ by (now indicating the m-dependence explicitly)

$$S^m_{jj'} = \langle jm|e^{2i\eta(\theta)}|j'm\rangle \qquad (14)$$

The classical result is obtained as the stationary phase-approximation of (14).The stationary phase angle θ_0 for large j,j' is given by

$$j'-j = 2(\partial\eta(\theta)/\partial\theta)_{\theta=\theta_0} = \Delta j(\theta_0) \qquad (15)$$

and for the intensity, one finds expanding the exponent of (14) around θ_0 up to second order

$$P(\Delta j) = \text{const}\ |d\theta_0(\Delta j)/d\Delta j| \qquad (16)$$

So whenever the classical "excitation function" $\Delta j(\theta)$ is stationary with respect to the reference angle θ one obtains a singularity in the classical intensity distribution (16).

In the so-called "sudden" approximation the phase-shift is approximated by the WKB integral [10,11]

$$\eta(\theta) = \lim_{z\to\infty} \{\int_{z_0}^{z} (k^2-v(\theta,z))^{1/2}dz - kz\} \ ; \ z_0 = \text{classical turning point} \qquad (17)$$

leading to an excitation function

$$\Delta j(\theta) = -kv'/v - d(v'/v - 2w'/w)\ (\pi/2+ \arcsin(d/k_d)) \qquad (18)$$

where $d=w/(2\sqrt{v})$ and $k_d^2 = k^2+d^2$. The prime (') indicates differentiation with respect to θ. In the simplest version of the sudden approximation k is the wave number corresponding to the total energy (see (2)), an improvement consists of using an average wave number given by $2k=k_j+k_{j'}$, for the transition from j to j'. This prescription, however, does not guarantee the unitarity of the S-matrix (14) any more. The physical basis of the sudden approximation is the short time scale of the translational impact which allows to neglect the rotation of the molecule during the collision. The phase-shift (17) can also be used for the semiclassical S-matrix (14). This "quantum sudden" approximation yields rather good intensities for low rotational energy-transfers. Since the average energy-transfer

$$\Delta\varepsilon_j = \sum_{j'} rj'(j'+1)|S_{jj'}|^2 - rj(j+1) \qquad (19)$$

is only of the order of 5% to 20% of the incident kinetic energy k_j^2 the quantum sudden approximation even yields acceptable overall results for the intensity distributions (see Fig.1). For repulsive potentials the general tendency is to overestimate the position of the high-energy rainbow and the average energy-transfer somewhat. This is due to the assumed orientational rigidity during the collision. Relaxation of this rigidity usually reduces the torque on the molecule during the collision leading to a smaller energy-transfer. The situation is more complicated for potentials with attractive wells. In this case the sudden approximation can lead to positive as well as negative deviations of the rainbow positions from their exact values. This is indicated in Fig.2 which exhibits classical excitation functions calculated from (18) as compared to exact classical trajectory results.

The similarity between the two excitation functions of Fig.2 suggests a simple improvement of the quantum sudden approximation: One multiplies the r.

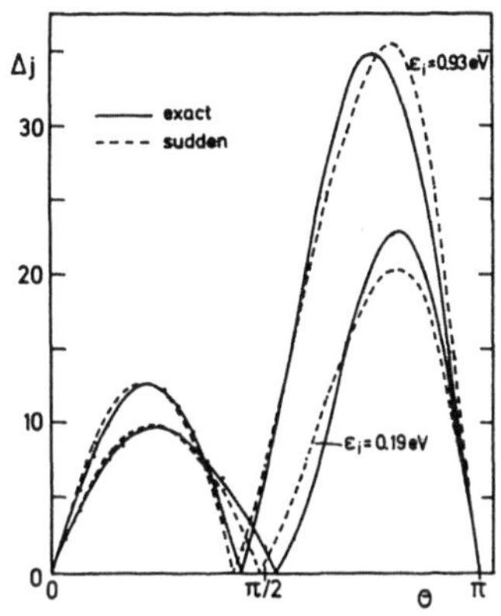

Fig.2 Classical excitation functions for the same potential as used for Fig.5.

h.s. of (17,18) by a constant factor to shift the classical rainbow position to its correct value.Fig.1 contains an intensity distribution obtained from (14) using (18) with a factor 0.92.As one can see,this reproduces the exact quantum mechanical result rather well.

Figure 1 also contains the classical rainbow singularities calculated from (16) using the exact classical excitation function (see Fig.2).The classical intensity distributions are much too narrow near the rainbow maxima.This situation is similar to the vibrational case where quantum effects of the phonon system also lead to considerable broadening of the energy distribution of scattered particles [14].Generally speaking the quantum distributions leak out into the classically "forbidden" regions above the classical rainbow.Quantum theory thus leads to somewhat larger average energy-transfers than the classical approximation.Figure 3 shows the comparison of calculated average energy transfers as a function of incident energy for potentials with and without attractive well and for various approximations.

Particularly interesting is the dependence of the average energy-transfer on the initial rotational quantum number j. One might expect that, for fixed total energy,there will be a critical initial j above which energy,on the average,will be transferred from rotation to translation. A glance at Fig. 4 shows that,indeed,this is what happens.There are some details which are associated with the presence or absence of an attractive well. But in both cases there exists some critical initial rotational energy E_c where the average energy-transfer changes sign.

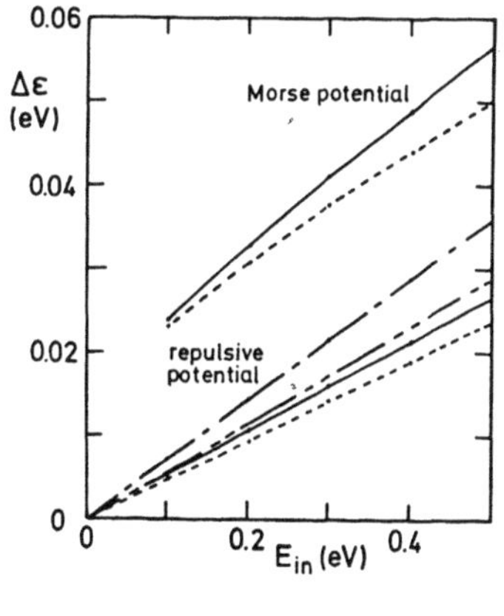

Fig.3 Average energy-transfer for a Morse potential $V = 1+0.21P_1(\cos\theta)+0.14P_2(\cos\theta)$, w=38.4, α = 1.4/Å and the repulsive potential of Fig.1.

———————: Quantum theory
----------: Classical theory
—— - ———: Quantum sudden approximation
—— -- ——: Classical sudden approximation

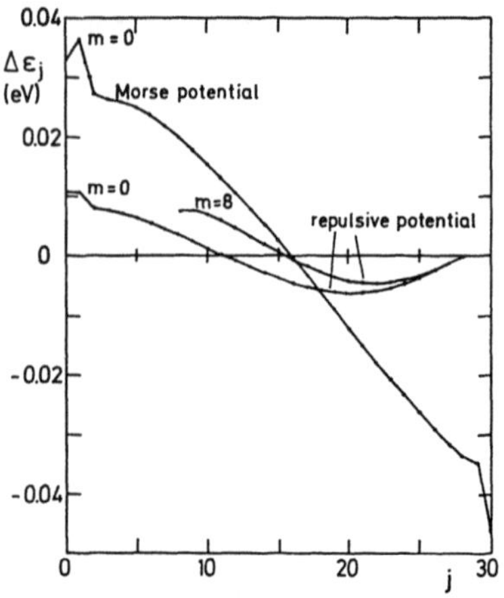

Fig.4 The average energy-transfer as a function of initial j value, for an incident energy E=0.2 eV. The Morse potential is the same as in Fig.3, the repulsive potential the same as in Fig.1. For the repulsive potential the m-dependence is also shown.

This result can easily explain the experimental findings [5] on the rotational cooling of desorbing molecules in a semiquantitative way.As is well known from kinetic theory[16] the energy distribution of desorbing particles is a Boltzmann distribution with the surface temperature T_s multiplied by the sticking coefficient s(E)

$$P(E) = \text{const}\ s(E)\ \exp(-E/kT_s) \tag{20}$$

The linear dependence of lnP(E) on E obtained experimentally indicates an exponential dependence $s(E) = \exp(-E/E_0)$ of s leading to an effective temperature of the desorbing molecules $T_{eff} = T_s E_0/(kT_s+E_0)$.It is now very tempting to identify E_0 approximately with the critical value E_c from above: Slightly above E_c one expects the negative energy-transfer to rotations to overcompensate the positive one to the phonons.Since this prevents the molecules from sticking one expects a sharp drop of s(E) above E_c. Although this simple argument can not replace a decent calculation of s(E) it is gratifying to note that the value of $E_c \cong 400$K found in Fig.4 agrees rather well with the saturation values of T_{eff} found experimentally [5].

In order to fit the experimental rainbow intensities it seems to be important to use a potential with an attractive well of appropriate depth and angular dependence w(θ).Figure 5 exhibits an intensity distribution with smaller intensity in the high energy rainbow region than Fig.1 obtained using a purely repulsive potential.

The intensities of Figs.1 and 5 correspond to an initial angular momentum j=0.Experimentally one can not realize this condition.One rather has some initial rotational temperature T_i= 10K to 50K. Figure 6 shows results of a final state distribution averaged over an initial Boltzmann distribution

$$P_{j'}(T_i) = \sum_j |S^m_{jj'}|^2 \exp\{-\hbar^2 j(j+1)/(2IkT_i)\}/Z(T_i) \tag{21}$$

where $Z(T_i)$ is the rotational partition function.As one can see the main effect of the non-zero initial temperature is to average out the oscillations of the scattering intensity (13) as shown in Figs.1,5.The average intensity (21) only exhibits two smooth peaks corresponding to the two rainbows.

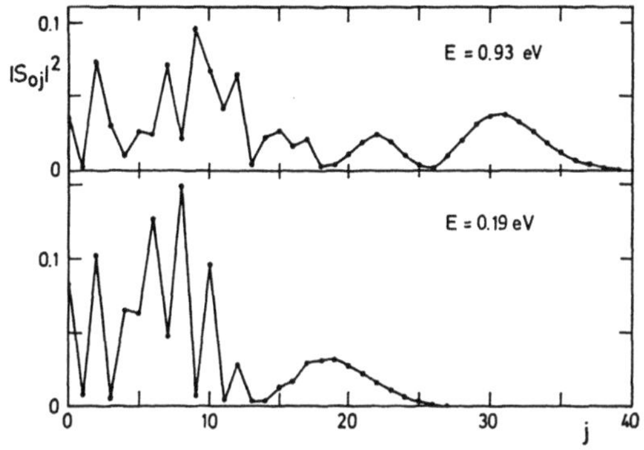

Fig.5 Rotational intensities from coupled-channel calculations for a Morse potential $v = 1+0.19\cos\theta-0.25\cos^2\theta$, $w = 58+7.42\cos\theta-5.88\cos^2\theta$, $\alpha=2/\text{Å}$, for an initial rotational temperature $T_i=0$, two different incident energies.

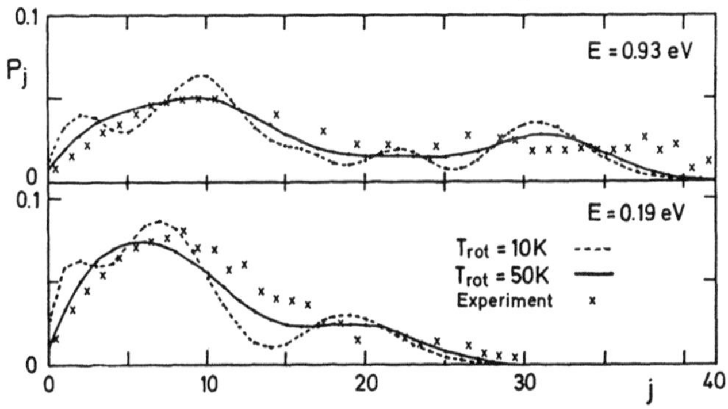

Fig.6 Rotational intensities for the same input data as in Fig.5 but averaged over an initial rotational distribution (see(21)) with $T_i=T_{rot}$. Quantum structure which is still visible for T_{rot}=10K is wiped out completely for 50K.

It may be of interest to mention that the "smoothing effect" of an initial rotational temperature on the final state distribution is much more efficient than the additional broadening due to a non-zero surface temperature T_s via the inelasticities due to the excitation of substrate phonons which we are going to consider in the next section.

4. Effect of substrate vibrations

If the energy-transfer to the substrate phonons becomes large compared to the Debye energy, the corresponding energy distribution becomes insensitive to the phonon spectrum (see BÖHEIM,BRENIG [14]).For a first orientation it is therefore sufficient to consider an Einstein model for the substrate phonons.We choose the Einstein frequency equal to the average frequency $\overline{\omega}=3\omega_D/4$ of a Debye spectrum.If we use the same dimensionless units as introduced in (1),(2) the translational wave number is now given by

$$k^2_{\ell,\mu} = k^2 - r\ell(\ell+1) - \omega(\mu+1/2). \tag{22}$$

Here ω is the dimensionless Einstein frequency given by

$$\omega = \overline{\omega}\, 2M/(\hbar\alpha^2) \tag{23}$$

and μ is the oscillator quantum number. In addition to the range constant α we introduce the reciprocal length unit α_s for the substrate oscillators by

$$\alpha_s^2 = M_s\overline{\omega} / \hbar \tag{24}$$

and denote the dimensionless displacement coordinate of the oscillator by x. The strength functions v,w of the Morse potential (9) will then be functions of x and θ.For the sake of simplicity we consider only repulsive potentials with a P_2-anisotropy

$$v(x,\theta) = \exp(2\alpha x/\alpha_s)(1+\lambda P_2(\cos\theta)) \tag{25}$$

In analogy to (10) the Jost functions are given as a power series

$$f_{\ell;\mu'}(\ell,\mu,z) = \sum_n f_{\ell;\mu'}(\ell,\mu,n)\, e^{-(n+ik_{\ell;\mu'})z} \tag{26}$$

and the recurrence relation (11) reads

$$f_{\ell;\mu'}(\ell,\mu,n) = \frac{1}{(n+ik_{\ell;\mu'})^2 + k_{\ell,\mu}^2}\, v\, f_{\ell;\mu'}(\ell,\mu,n-2) \tag{27}$$

As a check of approximations we first computed vibrational intensities

$$P_{\mu,\mu'} = |S_{\mu,\mu'}|^2 \tag{28}$$

neglecting rotational transitions altogether. Fig.7 shows the exact results for $P_\mu = P_{0,\mu}$ using (27) for a purely repulsive potential with $\alpha=2/\text{Å}$, a Debye temperature of 225K and an incident energy of 0.2eV as compared to the sudden approximation and a Gaussian (denoted BB) whose position and width was calculated from eqs. (17,17a) of BÖHEIM and BRENIG [14].In this approximation,in contrast to the sudden approximation,the unperturbed vibrational motion during the collision is taken into account;only the perturbation of this motion by the collision is neglected. While this approximation works quite well,the sudden approximation overestimates the vibrational effects strongly.

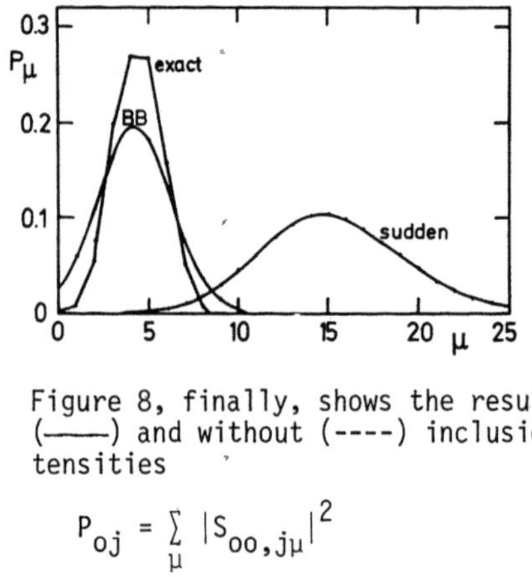

Fig.7 Vibrational intensities from coupled channel calculations for NO/Ag as compared to BB and sudden approximation (see text above).

Figure 8, finally, shows the results of computed rotational intensities with (——) and without (----) inclusion of phonons.It can be seen that total intensities

$$P_{oj} = \sum_\mu |S_{oo,j\mu}|^2 \tag{29}$$

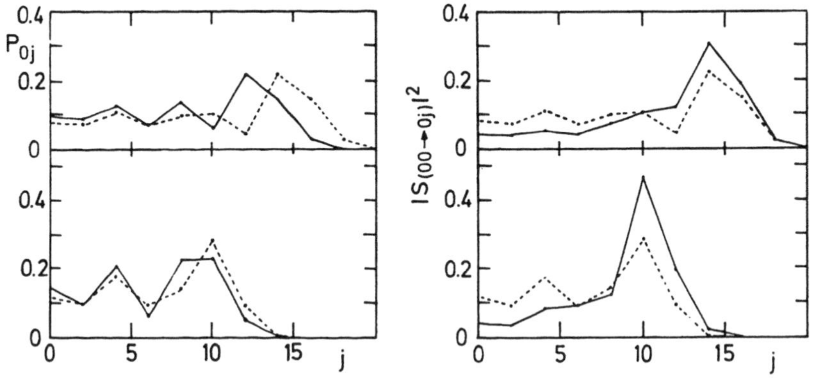

Fig.8 Total rotational intensities from coupled channel calculations (left) as compared to Debye-Waller factors renormalized to 1 (right) for incident energy of 0.2eV (upper part) and 0.1eV (lower part) see text above, λ=0.6 see eq. (25).

are shifted to lower rotational energies and broadened somewhat by the vibrational transitions,while the zero phonon intensities are shifted to higher rotational energies and sharpened slightly.The influence of vibrational transitions on the rotational intensities altogether,however,does not appear to be very large as compared to the quantum effects.

References

1. R.G.Rowe and G.Ehrlich,J.Chem.Phys.63,4648(1975); G.Boato,P.Cantini and L.Mattera,ibid.65,544(1976); J.P.Cowin,C.F.Yu,S.J.Sibener and J.Hurst,J.Chem.Phys.75,1033(1981)
2. W.Allison and B.Feuerbacher,Phys.Rev.Lett.45,2040(1980)
3. F.Frenkel,J.Häger,W.Krieger,H.Walther,C.T.Campbell,G.Ertl,H.Kuipers and J.Segner,Phys.Rev.Lett.46,152(1981)
4. A.W.Kleyn,A.C.Luntz and D.J.Auerbach,Phys.Rev.Lett.47,1169(1981)
5. J.Segner,H.Robota,W.Vielhaber,G.Ertl,F.Frenkel,J.Häger,W.Krieger and H.Walther, Surface Sci.131,273(1983)
6. G.D.Kubiak,J.E.Hurst Jr.,H.G.Rennagel,G.M.McClelland and R.N.Zare, J.Chem.Phys.79,5163(1983)
7. C.F.Yu,C.S.Hogg,J.P.Cowin,K.B.Whaley,J.C.Light and S.J.Sibener,Israel J.Chem.22,305(1982); R.Schinke,Surface Sci.127,283(1983)
8. J.A.Barker,A.W.Kleyn and D.J.Auerbach,Chem.Phys.Lett.97,9(1983)
9. H.Voges and R.Schinke,Chem.Phys.Lett.100,245(1983)
10. R.B.Gerber,A.T.Yinnon,Y.Shimoni and D.J.Kouri,J.Chem.Phys.73,4397(1980)
11. H.Voges and R.Schinke,Chem.Phys.Lett.95,221(1983)
12. S.Tanaka and S.Sugano,Surface Sci.136,488(1984)

13. a)J.E.Hurst Jr.,G.D.Kubiak and R.N.Zare,Chem.Phys.Lett.93,235(1982);
S.Tanaka and S.Sugano,Surface Sci.143,L371(1984)
b)R.Schinke and R.B.Gerber,preprint (1984)

14. J.Böheim and W.Brenig,Z.Phys.B41,243(1981);
H.D.Meyer,Surface Sci.104,117(1981)

15. R.Jost and A.Pais,Phys.Rev.82,840(1951)

16. W.Brenig,Z.Phys.B48,127(1982);
U.Leuthäusser,Z.Phys.B50,65(1983)

Dynamical Aspects in Chemical Reactions Near Solid Surfaces
- Special Reference to the Ortho-Para Conversion on Magnetic Surfaces -

Y. Ishii

The Institute for Solid State Physics, The University of Tokyo
Roppongi, Minato-ku, Tokyo 106, Japan

1. Introduction

In studying chemical reactions which occur near solid surfaces, we must consider several time constants: namely, an intrinsic time constant τ_{int} which corresponds to an inverse of the energy released from a chemical species by the reaction, a surface life time τ_{life} for which a chemical species stays near solid surfaces and suffers a perturbation from them, a response time τ_{res} which a substrate requires to respond to a perturbation and so on. Especially the response time of a substrate is related to a relaxation rate of the heat of reaction and so the dynamical response of the substrate may have a remarkable effect which cannot be seen in homogeneous catalytic reaction[1].

If the chemical species has a contact with the perturbation source for a long enough time compared with the other time constants such as τ_{int} and τ_{res}, the quasi-stationary approximation is justified for the calculation of the reaction-rate. That is, the transition-rate based on the Fermi's golden rule is regarded as representing a time constant for the reaction. If the surface life time is not so long, however, we must go beyond the quasi-stationary approximation. In §2, we examine the transient behaviour in the population change of the final state within a simplified model and discuss the competition between the surface life time and the other time constants.

When the quasi-stationary approximation is valid, the dynamical response of the substrate is fully reflected to the reaction rate. In §3, we investigate the ortho-para conversion of a hydrogen molecule on magnetic surfaces which is one of the typical magnetocatalytic reactions. For an H_2 molecule, the nuclear spin must form a triplet(singlet) in the spatial-coordinate state of the odd(even) parity for the space inversion because of Pauli's principle. In the ortho-para conversion on various magnetic insulators[2], a strongly inhomogeneous dipole field coming from magnetic moments due to electronic spins in the substrate causes simultaneous excitations of the proton spins between a triplet and a singlet states and the rotation of an H_2 molecule[3-11].

2. Transient Behaviour in Surface Reactions

In order to take into account the surface life time explicitly, we adopt the trajectory approximation. The motion of a chemical species is assumed to be given by a classical trajectory and then the position-dependent Hamiltonian is converted into the time-dependent one[12]. For simplicity we consider a two-level system as the chemical species and assume the boson system as the substrate degrees of freedom. The model

Hamiltonian is given by

$$H(t) = H_c + H_s + V(t), \tag{2.1a}$$

$$H_c = \frac{1}{2}\varepsilon_0[|1\rangle\langle 1| - |2\rangle\langle 2|], \tag{2.1b}$$

$$H_s = \sum_q \hbar\,\omega_q\, b_q^\dagger b_q\,, \tag{2.1c}$$

$$V(t) = f(t)\sum_q V_q(b_q + b_q^\dagger)[|1\rangle\langle 2| + |2\rangle\langle 1|], \tag{2.1d}$$

where $|i\rangle$ (i=1,2) describes the eigenstate of the chemical species and the energy difference between them ε_0 is assumed to be constant. The boson system is described by the annihilation (creation) operator b_q ($b_q^\dagger$) with the frequency ω_q for an appropriate quantum number q. The explicit time-dependence $f(t)$ is assumed to be a rectanguler form with duration time τ_{life} as shown in Fig.1.

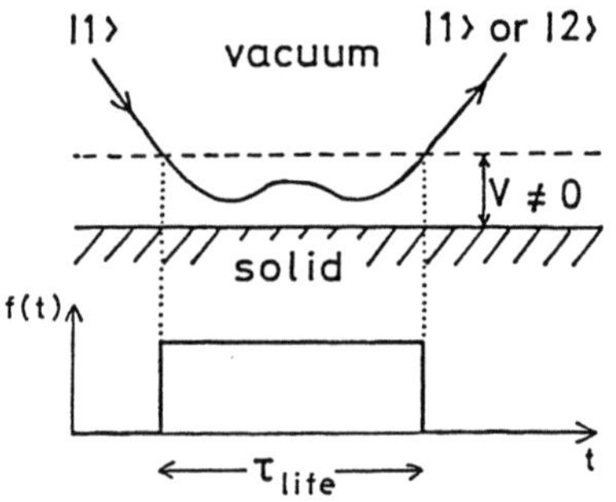

Fig. 1 Schematic illustration for surface reactions. The perturbation operates to a chemical species during a finite time τ_{life}.

The time evolution of the total system is described by the equation of motion for the density matrix $\rho(t)$

$$i\hbar\frac{\partial\rho}{\partial t} = [H(t), \rho(t)], \tag{2.2}$$

We assume that the substrate is in its equilibrium before the perturbation is switched-on and the species coming to the surface is in the $|1\rangle$ state as

$$\rho(t_0) = |1\rangle \exp(-\beta H_S)\langle 1|,$$

where $\beta = 1/k_BT$. The population of the $|i\rangle$ state is calculated as

$$N_i(t) = Tr^{(c,s)}[|i\rangle\langle i|\,\rho(t)] = Tr^{(s)}\langle i|\rho(t)|i\rangle, \tag{2.3}$$

where $Tr^{(c,s)}$ is the trace over the states of the chemical species(c) and the substrate degrees of freedom(s). If we confine our discussion to the first-order perturbation theory, the population $N_i(t)$ is calculated as

$$N_1(t) = 1 - F^{(2)}(t) \cong \exp[-F^{(2)}(t)], \tag{2.4a}$$

$$N_2(t) = 1 - N_1(t), \tag{2.4b}$$

with

$$F^{(2)}(t) = \int_{t_0}^{t} ds \int_{t_0}^{t} ds' \langle\langle 1|V_I(s)V_I(s')|1\rangle\rangle_S, \tag{2.5}$$

where $V_I(t)$ is the interaction representation of $V(t)$, $<...>_s$ means the thermal average for the substrate system and we use the units of $\hbar=1$. Assuming that the spectrum for the substrate boson is continuous, we obtain the population at $t=\infty$, $N_1(\infty)$ in terms of $F^{(2)}(\infty)$ calculated as

$$F^{(2)}(\infty) = \int_0^\infty d\omega <V_q^2>_{\omega_q=\omega} D(\omega)[n(\omega)|\hat{f}(\varepsilon_0+\omega)|^2 + (n(\omega)+1)|\hat{f}(\varepsilon_0-\omega)|^2]. \tag{2.6}$$

with

$$\hat{f}(\omega) = \int_{-\infty}^{\infty} dt \exp(i\omega t) f(t),$$

where $D(\omega)$ is the density of state for the substrate boson and $n(\omega)=[\exp(\beta\omega)-1]^{-1}$. To be more precise, we consider the spectrum of the surface Debye phonon to be

$$\begin{aligned} D(\omega) &= 2\omega/\omega_D^2, \quad & 0<\omega<\omega_D \\ &= 0, & \text{otherwise} \end{aligned} \tag{2.7}$$

with the Debye frequency ω_D and the completely local interaction to be

$$V_q = V. \quad \text{(const.)} \tag{2.8}$$

Finally we obtain the expression for $F^{(2)}(\infty)$ as

$$F^{(2)}(\infty) = \left(\frac{2V}{\omega_D}\right)^2 \int_{-\omega_D}^{\omega_D} d\omega \frac{\omega \exp(\beta\omega)}{\exp(\beta\omega)-1} \frac{1-\cos(\varepsilon_0-\omega)\tau_{life}}{(\varepsilon_0-\omega)^2}, \tag{2.9}$$

If the two-level system is perturbed by the off-diagonal mixing V, the population of each level fluctuates. This fluctuation due to the quantum origin has a frequency of the order of ε_0. The annihilation and creation of the boson act as the energy-transfer to and from the two-level system. The real energy-transfer, however, occurs with a time delay corresponding to the response time of the substrate and the resonance condition between the response of the substrate and the fluctuation of the perturbed two-level system must be satisfied. In order to see these features, we examine the dominant τ_{life}-dependence in the several time range at T=0 (Table 1).

In the range of I, the substrate cannot follow the perturbation so that the population change does not depend on the physical quantity relevant to the substrate, ω_D. In II, the substrate can respond to the perturbation but

Table 1 The most dominant τ_{life}-dependence of $F^{(2)}(\infty)$ at T=0.

range	τ_{life}	$F^{(2)}(\infty)$
I	$\tau_{life} \ll 1/\varepsilon_0,\ 1/\omega_D$	$(V\tau_{life})^2$
II	$1/\omega_D \ll \tau_{life} \ll 1/\varepsilon_0$	$(\frac{2V}{\omega_D})^2 \ln \omega_D\tau_{life}$
III	$1/\omega_D < 1/\varepsilon_0 \ll \tau_{life}$	$2\pi V^2 D(\varepsilon_0)\ \tau_{life}$
II'	$1/\varepsilon_0 \ll \tau_{life} \ll 1/\omega_D$	$2(V/\varepsilon_0)^2(1-\cos\varepsilon_0\tau_{life})$
III'	$1/\varepsilon_0 < 1/\omega_D \ll \tau_{life}$	$2(V/\varepsilon_0)^2(1+\sin\varepsilon_0\tau_{life}/2\omega_D\tau_{life})$

does not resonate with the fluctuation in the species. The logarithmic dependence in this range comes from the low-frequency response of the substrate. If the species stays on the surface so long that the substrate can fully follow the perturbation and its response completely resonates with the fluctuation of the species (the range III), the population change is given by $\exp(-\Gamma\tau_{life})$ where the decay rate Γ is calculated by using the Fermi's golden rule as

$$\Gamma = 2\pi V^2 D(\varepsilon_0). \tag{2.10}$$

It should be noted that the higher order perturbation must be included if $\Gamma\tau_{life}>1$. If $\omega_D<\varepsilon_0$, there is the time range (II') where the substrate cannot follow the fluctuation of the species. Then, the final state population is described by the fluctuation of the perturbed two-level system. In III', $F^{(2)}(\infty)$ tends slowly to a finite value as $\tau_{life}\to\infty$. Apparently this behaviour differs from that predicted by the golden rule, $F^{(2)}(\infty)=0$. This is due to the nonadiabatic switch-on/off of the perturbation assumed in the present model.

3. Ortho-Para Conversion on Magnetic Surfaces[10,11]

In the preceding section, we discussed the competition between the surface life time and the other time constants. The surface life time of a hydrogen molecule is estimated as 10^{-12} sec [3,13] at room temperatures while τ_{int} and τ_{res} are of the order of $10^{-14}\sim 10^{-13}$ sec. Therefore we can adopt the quasi-stationary approximation for the ortho-para conversion. That is, we neglect details of adsorption and desorption processes and assume that the molecule moves parallel to the surface at the distance of Z_0. The ortho-para mixing V_{op} originates from the inhomogeneous dipole field due to the electronic spins in the substrate.

The transition probability is calculated within the lowest order perturbation theory as

$$W = \int_{-\infty}^{\infty} dt \langle\langle i|V_{op}(t)|f\rangle\langle f|V_{op}(0)|i\rangle\rangle_{st} , \tag{3.1}$$

where $V_{op}(t)$ is the interaction representation of the ortho-para mixing and $\langle\ldots\rangle_{st}$ means a statistical average over the translational motion of the molecule and the motion of the substrate spins. The initial and the final states are identified by a set of variables (ℓ, m, I, I_z) where ℓ and m are the rotational quantum numbers and I and I_z are the total nuclear spin and its z-component. For simplicity, we consider only spins in the surface layer:

$$S_q = \frac{1}{N}\sum_{i=1}^{N}\exp(-iq\,R_i)\,S_i, \tag{3.2}$$

where R_i, S_i, N and q are the position coordinate, the spin operator of the substrate atom at the i-th site, the total number of spins in the surface layer and the two dimensional(2D) wavevector, respectively.

The transition probability from (ℓ, m, I, I_z) to (ℓ', m', I', I_z') is obtained as

$$W = \left(\frac{\pi\mu_e\mu_p}{2a^3}\right)^2 \frac{b^2}{9a^2}(\ell+\ell'+1)f(\ell m I I_z;\ell' m' I' I_z')(W_B+W_D). \tag{3.3}$$

$$W_B = \langle S_Q^z\rangle^2 \sum_G |a(Q+G)|^4 \exp(-2|Q+G|Z_0)\Lambda_{Q+G}^{(mol)}(\varepsilon_0), \tag{3.4a}$$

$$W_D = \sum_q |a_q|^4 \exp(-2|q|Z_0)\int_{-\infty}^{\infty}\frac{d\omega}{2\pi}\Lambda_q^{(mol)}(\varepsilon_0-\omega)[\Lambda_q^{(zz)}(\omega)+\Lambda_q^{(xx)}(\omega)], \tag{3.4b}$$

where a and b are the lattice constant of the substrate and the bond length of a hydrogen molecule, respectively. The definition of $f(\ell m I I_z;\ell' m' I' I_z')$ is given in ref.[10]. The summation of q is over the whole of the 2D reciprocal lattice space and Q and G are the wavevector giving the non-vanishing $\langle S_Q^z \rangle$ below the magnetic critical temperature T_c and the 2D reciprocal lattice vector, respectively. The heat of reaction ε_0 is the energy released by the $(\ell, m, I, I_z) \to (\ell', m', I', I_z)$ conversion. The spectral functions are defined as

$$\Lambda_q^{(mol)}(\omega) = \int_{-\infty}^{\infty} dt\, \exp(i\omega t) \langle \exp[iq(R_{/\!/}(t) - R_{/\!/}(0)] \rangle, \tag{3.5}$$

$$\Lambda_q^{(zz)}(\omega) = \int_{-\infty}^{\infty} dt\, \exp(i\omega t) \langle \Delta S_q^z(t)\ \Delta S_{-q}^z(0) \rangle, \tag{3.6a}$$

$$\Lambda_q^{(xx)}(\omega) = \int_{-\infty}^{\infty} dt\, \exp(i\omega t) [\langle S_q^x(t) S_{-q}^x(0) \rangle + \langle S_q^y(t) S_{-q}^y(0) \rangle], \tag{3.6b}$$

where $R_{/\!/}(t)$ is the lateral position of the molecule at time t and

$$\Delta S_q^z = S_q^z - \delta_{q,Q} \langle S_Q^z \rangle .$$

In eq.(3.3), W_B and W_D correspond, respectively, to the Bragg scattering and the diffuse scattering in the neutron magnetic scattering. For the Bragg scattering part, the heat of reaction is not transferred to the substrate. On the other hand, for the diffuse scattering part, a part of the heat of reaction, ω, is transferred to the substrate and the remaining part, $\varepsilon_0 - \omega$, to the translational motion. These two possible paths of energy-flow cooperate to determine the conversion rate as expressed in terms of the convolution of two spectral functions.

As for the substrate spin system, we take into account only spins in the surface layer. If the solid surface is treated as the 2D system, we are faced with several unsatisfactory features inherent in the low dimensional system. In this paper, we first calculate the spectral function for the three dimensional spin system and then obtain the relevant spectral function in eq.(3.6) by summing over the q_z component. The summation of q in eq.(3.4b) is taken over the first Brillouin zone of the substrate lattice because the contribution from the larger q is rather small due to the factor of $\exp(-2|q|Z_0)$. Therefore the diffuse scattering part is calculated by using the formulae

$$W_D = \sum_{k \text{ in BZ}} [\mathcal{L}_k \langle \Delta S_k^z \Delta S_{-k}^z \rangle + \mathcal{T}_k (\langle S_k^x S_{-k}^x \rangle + \langle S_k^y S_{-k}^y \rangle)], \tag{3.7}$$

where

$$\mathcal{L}_k = |aq|^4 \exp(-2|q|Z_0) \int_{-\infty}^{\infty} \frac{d\omega}{2\pi} \Lambda_q^{(mol)}(\varepsilon_0 - \omega)\, \Lambda_k^{(zz)}(\omega) / \langle \Delta S_k^z \Delta S_{-k}^z \rangle, \tag{3.8a}$$

$$\mathcal{T}_k = |aq|^4 \exp(-2|q|Z_0) \int_{-\infty}^{\infty} \frac{d\omega}{2\pi} \Lambda_q^{(mol)}(\varepsilon_0 - \omega)\, \Lambda_k^{(xx)}(\omega) / (\langle S_k^x S_{-k}^x \rangle + \langle S_k^y S_{-k}^y \rangle), \tag{3.8b}$$

and k denotes (q, q_z). The convolution integrals $\mathcal{L}_k$ and $\mathcal{T}_k$ describe the ability of the relaxation of the heat of reaction while the equal-time correlation $\langle S_k^\alpha S_{-k}^\alpha \rangle$ the magnitude of the perturbation.

The microscopic calculation of the spectral functions is very difficult especially for a short wavelength or a high frequency. We assume a simple functional form. The diffuse scattering parts of the transition probability are shown in Figs.2 for T_c=300 (K) and ε_0=172 (K) which is the energy difference between ℓ=0 and 1 for a free hydrogen molecule. In Figs.2, τ^* is a parameter for the spectral functions of the spin system which becomes broader for large τ^*. We find a qualitatively similar temperature dependence of W_D for ferromagnetic and antiferromagnetic cases. The quantitative difference comes from the fact that, in an antiferromagnet, the spin fluctuation with a short wavelength has a large amplitude and produces a field of large inhomogeneity. The dependence upon τ^* is rather conspicuous[11]. For the case of the narrow spectrum ($\tau^*=0.5$), the enhancement of W_D below T_c is caused by that of $\mathcal{T}_k$ due to the excitation of the collective mode. On the other hand, for the case of $\tau^*=1.5$, the transition probability is determined in a more complicated way. The equal-time correlation $\langle S_k^\alpha S_{-k}^\alpha \rangle$ competes with the overlap integrals in determining the temperature-dependence of W_D.

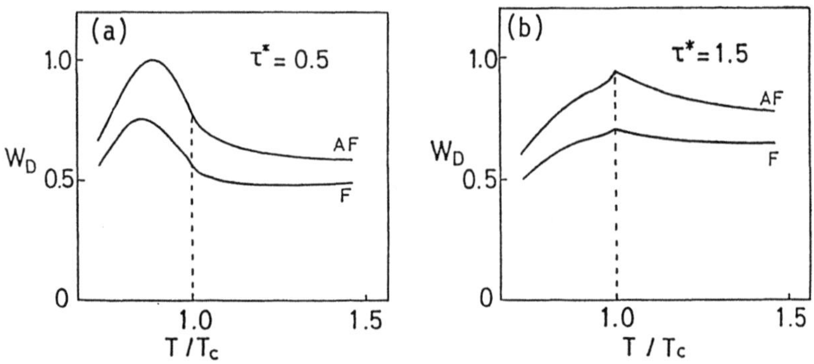

Fig. 2 Temperature-dependence of W_D for (a), $\tau^*=0.5$; and (b), $\tau^*=1.5$. The antiferromagnetic and ferromagnetic cases denoted by AF and F, respectively.

In the experimental result for Cr_2O_3[2], a drastic increase of the activity is observed at T_c when the sample is cooled through T_c. PETZINGER and SCALAPINO[7] ascribed this drastic change to the temperature-dependence of the magnitude of the perturbation $\langle S_k^\alpha S_{-k}^\alpha \rangle$ near T_c but the present authors[10] to the enhancement of the overlap integral $\mathcal{T}_k$ due to the collective mode excitation below T_c. At the present stage of experimental study, however, we cannot conclude which is dominant to determine the temperature-dependence of the conversion rate. More extensive studies such as measurements over the wider range of temperature on single crystal samples are strongly desired.

4. Summary

Dynamical aspects characteristic in heterogeneous catalytic reactions are discussed. (1) When a chemical species interacts with a substrate during a finite time τ_{life}, the final state population is not always given in terms of a constant decay rate as $\exp(-\Gamma\tau_{life})$ and the various τ_{life}-dependence is expected. (2) When the species interacts with a substrate for a long enough time, the dynamical response of the substrate is fully reflected to the reaction rate. In such a case, the energy flows to the relevant degrees of freedom cooperate in determining the activity. In both cases, the dynamical response of the substrate plays an important role. In order to discuss the chemical reactivity of solid surfaces in detail, further technical development may be needed for calculating dynamics of the substrate degrees of freedom.

References

1 E.G. d'Agliano, W.L. Shaich, P. Kumar and H. Suhl: Nobel Symposium 24, Collective Properties of Physical Systems (Academic Press, New York 1973), p. 200

2 P.W. Selwood: Adv. Catalysis 27, 23 (1978)

3 A. Leffler: J. Chem. Phys. 43, 4410 (1965)
4 E. Ilisca: Phys. Rev. Lett. 24, 797 (1970)
5 E. Ilisca and A.P. Legrand: Phys. Rev. B 5, 4994 (1972)
6 E. Ilisca and E. Gallais: Phys. Rev. B 6, 2858 (1972)
7 K.G. Petzinger and D.J. Scalapino: Phys. Rev. B 8, 266 (1973)
8 E. Ilisca: Phys. Rev. Lett. 40, 1535 (1980)
9 E. Ilisca, M. Debauche and J. L. Motchane: Phys. Rev. B 22, 687 (1980)
10 Y. Ishii and S. Sugano: Surf. Sci. 127, 21 (1983)
11 Y. Ishii and S. Sugano: J. Phys. C 17, 2049 (1984)
12 A. Blandin, A. Nourtier and D. Hone: J. Phys. (Paris) 37 369 (1976)
13 J.H. de Boer, The Dynamical Character of Adsorption (Oxford Univ. Press, New York 1953)

A Stochastic Trajectory Approach to Gas-Surface Inelastic Collisions

Shigeru Shindo

Department of Physics, Tokyo Gakugei University
Koganei, Tokyo 184, Japan

The optical potential method is introduced to deal with the inelastic scattering of gas atoms by solid surfaces. Using the local approximation for the optical potential, the semi-classical theories of the Debye-Waller factor and the energy-loss distribution are reexamined. A stochastic equation of motion taking into account the many phonon effect is applied to obtain the transition probabilities.

1 Introduction

Recent high quality experiments of scattering processes of atoms and molecules at solid surfaces, with which one obtains the important information about the surface interactions of gases, require the progress of the scattering theory for the experimental analysis. Furthermore the inelastic scattering of gases by surfaces is recognized as an important elementary process of the gas-surface reactions. Here after the term inelastic means the coupling of the atomic motion with acoustic phonon at surface, which gives the major contributions to the energy dissipation of atoms near surfaces. In the case where the heavier gases collide with surfaces, the effect of the many-phonon excitation cannot be neglected. However, we feel that there is no practical scattering theory which takes account of the many-phonon effect.

In this paper we propose to introduce the optical potential to atom-surface inelastic scattering, which allows us to take account of the multiple inelastic scattering effects. The utilization of the optical potential has been presented in the theories describing various scattering phenomena by many-body systems such as nuclei, atoms and solids [1]. Our present assertion is that the optical potential method is also powerful for the inelastic scattering of atoms by surfaces.

First we introduce the local approximation for the optical potential, which is based on the assumption that the wave packet state of the incident atom is localized. The attenuation of the elastic intensity due to the inelastic scattering is represented by the Debye-Waller factor. Describing the motion of the wave packet by using the classical approximation, we calculate the Debye Waller factor and the energy-loss distribution [2-6]. We propose here to represent the path of the atom by the use of the classical mechanical stochastic equation of motion including the effect of many phonon term.

2 Local Approximation of the Optical Potential

We start from the Schrödinger equation for scattering atoms in the time-dependent form as follows:

$$i\hbar\dot{\phi}_k(t) = (K + V)\ \phi_k(t) + \int^t dt'\ U(t,t')\ \phi_k(t') \qquad (1)$$

where U(t,t') is the optical potential expressed in the limit of the one phonon approximation by

$$U(t,t') = (1/i\hbar)\ F \exp\{iH(t-t')/\hbar\}\ F\int \exp\{iw(t-t')\}J(w)\ dw \quad (2)$$

and

$$J(w) = (1+n(w))(\hbar/mw)\ \rho(w)$$

n(w) the average occupation number ,$\rho(w)$ the phonon density of state of the surface, m the mass of a solid atom ,K the kinetic energy of scattering atom, V the interaction potential between atom and surface and $F = -\nabla V$. By using the "local time approximation" , which allows one to ignore the time difference of $\phi_k(t)$ and $\phi_k(t')$, we can show that the intensity of elastically scattered wave at time t, I(t), follows the differential equation

$$\dot{I}_{el}(t) = (2/\hbar)\int^t dt'\ \mathrm{Im}\langle U(t,t')\rangle_k\ I_{el}(t) \qquad (3)$$

where $\langle\ \ \rangle_k$ is $\langle\phi_k\ \ \phi_k\rangle$. Eq.(3) is solved as

$$I_{el}(t\to\infty) = \exp(-2W) \qquad (4)$$

where 2W is the Debye-Waller factor[3,5,6] expressed with the imaginary part of the optical potential by

$$2W = -(2/\hbar)\int dt\int^t dt'\ \ \mathrm{Im}\ \langle U(t,t')\rangle_k \qquad (5)$$

The Debye-Waller factor (5) is calculated from (2) as follows:

$$2W = \int dt\int dw\ \ P(w,E;t)$$

$$P(w,E;t) = (2/\hbar^2)\int^t \langle F(t)\ F(t')\rangle_k\ \cos\{w(t-t')\}\ J(w)\ dt' \qquad (6)$$

where P(w,E;t) may be interpreted as the transition probability per unit time at time t changing the energy of atom from E to $E - \hbar w$.

Now, we discuss the energy-loss distribution at time t , I(w,t) , using the classical multiple scattering method. The time variation of the energy-loss distribution I(w,t) of atoms obeys the master equation

$$\dot{I}(w,t) = \int P(w-w',E-w';t)\ I(w',t) - P(w',E-w;t)I(w,t)\ dw' \ (7)$$

In a simplified case where P(w,E;t) is independent of initial energies E, eq.(7) is solved with the initial condition $I(w,t=-\infty) = \delta(w)$, in the Fourier form as follows:

$$\tilde{I}(s,t\to\infty) = \exp[\int dt \int dw \quad P(w,E;t)\ \{\exp(iws)-1\}] \qquad (8)$$

where

$$I(w,t) = \int \tilde{I}(s,t)\ \exp(-iws)\ ds/(2\pi)$$

This result is identical to the result of many calculations [2-4] of the energy-loss distribution of the scattered atoms obtained with different methods. Moreover this agrees with the purely classical mechanical calculation[10] in the classical limit where the elastic scattering and the zero-point vibration of surface vanish. It should be noted that the master eq.(7) with the stationary (t-independent) transition probability is used to discuss the kinetics of adsorption and desorption of long time scale[8,9].

3 Self-Consistent Stochastic Trajectories

Now, we calculate the transition probability $P(w,E;t)$ using the trajectory method, where $F(t)$ is determined as a c-number function of t and $\langle F(t)\ F(t')\ \rangle_k$ in(6) is replaced by an

ensemble average over the trajectories $\langle F(t)F(t')\rangle$. Since F is an explicit function of the coordinate of atom z , our task is to obtain the classical atomic motion from eq.(1),which yields

$$\langle \dot{z}\rangle = \langle p\rangle/M\ ,$$

$$\langle \dot{p}\rangle = -\nabla V + \int^t dt' \int dw \langle\ \nabla F(t)F(t')\rangle\ J(w) \quad \sin\{w(t-t')\} \qquad (9)$$

and $F(t)$ in eq.(6) is $F(\langle z(t)\rangle)$. This is nothing but Kumamoto and Silbey's[4] result using the time-dependent Hartree approximation . As they point out, the mean energy-loss value calculated with eq.(8) is identical with the energy-loss of atoms obeying the equation (9), which is given by

$$\hbar\langle w\rangle = \hbar\int dw \int dt \quad w\ P(w,E;t)$$
$$= \int dw \int dt \int^t dt'\langle\ F(t)\ \ F(t')\rangle \cos\{w(t-t')\}\rho(w)/m \qquad (10)$$

We note here that eq.(9) is reduced to the Langevin equation when the atom-surface force can be approximated by the harmonic one

$$F(z(t')) = F(z(t)) - \{z(t) - z(t')\}\ \nabla F(z(t)) \qquad (11)$$

and by adding the "random force" F^R as follows:

$$\dot{z}(t) = p/M^*$$

$$\dot{p}(t) = -\nabla(V + V^{p.i}) - \eta p + F^R(t) \qquad (12)$$

where $V^{p.i.}$ and M^* are the phonon image potential and effective mass, respectively, both of which are given by

$$V^{p.i} = -(3/2mw_D{}^2)\ (\nabla V)^2$$

$$M^* = M(\ 1 - \Gamma)^{-1},\quad \Gamma = -(3/mMw_D{}^4)\ (\ \nabla^2 V)^2 \tag{13}$$

if the Debye spectrum is assumed. We note that large enhancement of the effective mass noticed by Knowles and Suhl

[7] does not occur in our perturbation calculation, but M^* is a little smaller than bare mass. And η the friction coefficient that is related to the random force with the fluctuation-dissipation theorem, is given by

$$\begin{aligned} M^*\eta &= (k_B T)^{-1} \int_0^\infty (F^R(0); F^R(t))\ dt \\ &= \langle\ \nabla F(z)^2\rangle\ 3\pi\ /(2m\ w_D{}^3) \end{aligned} \tag{14}$$

Till now, our treatment to determine $F(t)$ is limited to the one-phonon approximation where the coupling between the atom and the phonon is proportional to the displacement of phonon from the equilibrium position. Previously[10], we have proposed a classical mechanical model to take account of the many-phonon contribution to the trajectory of the atom. According to the model, the motion of the atom follows the equations of motion

$$M\ \ddot{z}(t) = F\{z(t) + \int^t dt'\ G(t-t')\ F(t') - Z^u\} = F(t) \tag{15}$$

,where Z^u represents the thermal motion of vibrating surface

atom and $k_B TG(t) = (\ \dot{Z}^u(t);\ \ Z^u(0))/m$. Eq.(15) is reduced to eqs.(9) or (12) when the displacement of the struck atom is negligible in comparison to the range of the potential. The mean energy-loss and the width of the energy-loss distribution calculated by using eq.(15) also coincide to those from the distribution (8) using the same $F(t)$. Then the distribution (8) calculated from eq.(15) ,which contains the many-phonon contribution, is the self-consistent energy loss distribution alternative to that calculated from eq.(9). In fig.(1), the Debye-Waller factor (5) is calculated as a function of the temperature, where $F(t)$ is obtained from (9), (15) and the elastic trajectory $M\ddot{z} = F(z)$. In fig.(2) the mean energy - loss value is shown as a function of incident energy. Fig.(3) the sticking probability defined by

$$s = \int_E^\infty dw\ I(w, t\to\infty) \tag{16}$$

is shown.

It is clarified that the many-phonon calculation of $F(t)$ with (15) reduces the inelastic scattering probability and that the one-phonon trajectory for $F(t)$ in (9) is wrong, especially for the calculation of the sticking probability. In fig.(2) and (3), E^{crit} is the critical trapping energy below which the deterministic trajectories of incident atoms given by (9) and (15) without Z^u are trapped by the surface potential. The sudden change of the mean energy-loss and the sticking probability at E^{crit} is smoothed out by the fluctuation of the

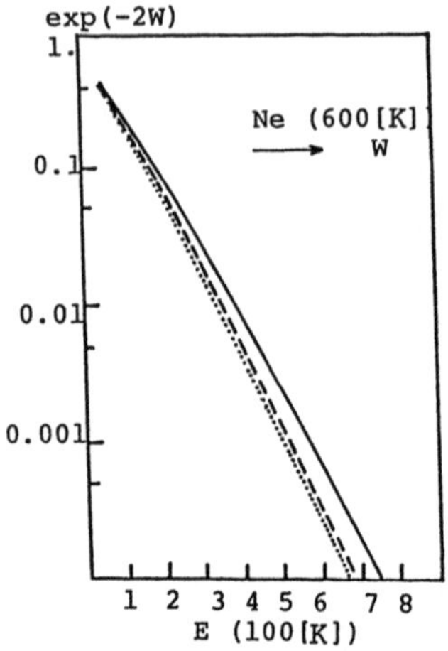

Figure(1):
Temperature - dependence of the Debye Waller factor exp(-2W) for a monoenergetic Ne atom colliding on W with the incident energy 600K. The parameter of the Morse potential, $V(z)=D(\exp(-2az)-\exp(-az))$, is

$a = 1.7\ A^{-1}$ and $D = 250K$, and the surface Debye temperature is chosen as 282K. Solid line is calculated with many phonon trajectory eq.(15), dotted line with one-phonon trajectory (9) and dashed line with elastic trajectory.

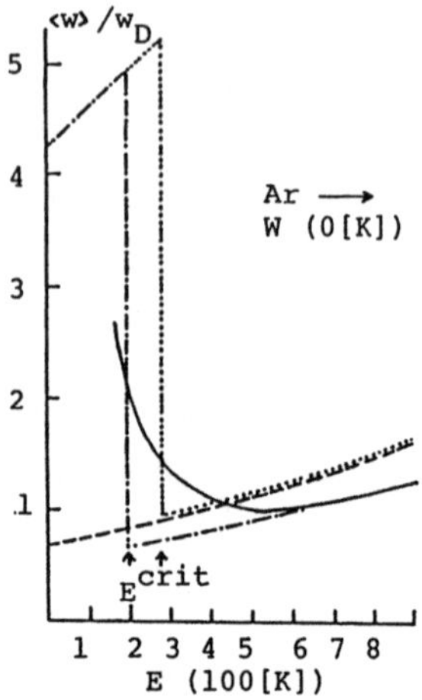

Figure(2):
Mean energy - loss as a function of the incident energy E. Ar atom collids to W, where

$a = 1.5\ A^{-1}$ and $D = 1200K$.
Temperature of solid is 0K. Solid line is calculated with many phonon trajectory eq.(15), dashed and dotted line with many phonon deterministic trajectory eq.(15) with $Z^u=0$, dotted line with one-phonon trajectory and dashed line with elastic trajectory.

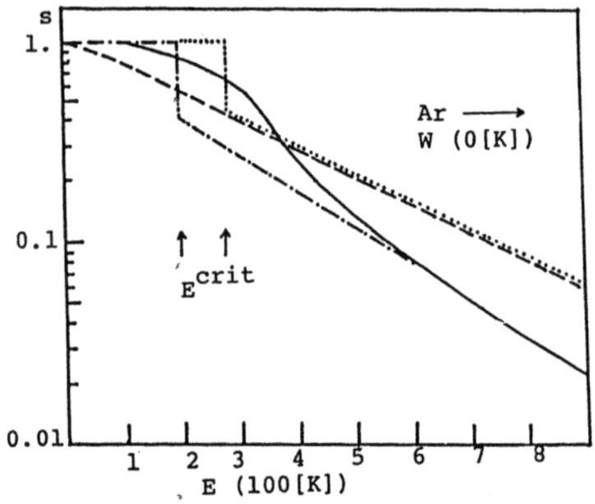

Figure(3):
Sticking probability s (16) as a function of incident energy. Each line has the same meaning with fig.(2).

struck atom Z^u. The utilization of eq.(8) for the energy-loss distribution is not appropriate below this energy. Instead the eq.(7) should be applied.

4 Concluding Remarks

Time-dependent optical potential theory developed here is an extention of the classical theory[10] to include the quantum effect, such as the recoilless elastic (Mossbauer) scattering and the zero-point vibration of surfaces. This is also the extension of the kinetic theory of gas-surface processes to the non-stationary case as eq.(7) shows. The many phonon trajectories are calculated from eq.(15) in order to obtain the transition probability in (6). It should be noted that the trajectory given by eq.(15) is unstable when the unharmonicity of atom-surface is important [10]. If the collision time exceeds several periods of atomic motion in the potential well, where the statistical correlation $\langle F(t)\ F(t')\rangle$ over the ensemble of trajectories depends only $(t-t')$, the collision process is treated in the framework of the kinetic theory of gases at surfaces[8,9]. The stochastic trajectories due to Z^u in eq.(15) play a dominant role in discussing the intermediate region of non-stationary and stationary transition probability in eq.(6)

The author is indebted to Professor S.Sugano for helpful discussions and critical reading of the manuscript.

References

[1]N.F.Mott and H.S.W.Massey:"The Theory of Atomic Collisions",Oxford University Press,Oxford (1965)
[2]J.Kaplan and E.Drauglis :Surf.Sci.36(1973)1
[3]W.Brenig :Z.Phys. B 36(1980) 81
[4]D.Kumamoto and R.Silbey :J.Chem.Phys. 75 (1981) 5164
[5]A.C.Levi and H.Suhl :Surf.Sci. 88 (1979) 221
[6]H.-D.Meyer :Surf.Sci. 104 (1981) 117
[7]T.P.Knowles and H.Suhl :Phys.Rev.Lett. 39 (1977) 1417
[8]U.Leuthausser :Z.Phys. B 44 (1981) 108
[9]W.Brenig :Z.Phys. B 48 (1982) 127
[10]S.Shindo :J.Phys.Soc.Jpn.52(1983) 562

Improvements to the Trajectory Approximation in Atom-Surface Scattering

D.M. Newns

Institute for Theoretical Physics, Santa Barbara, CA 93106, USA
and
Department of Mathematics, Imperial College, London SW7 2BZ, England

Abstract

We consider atom-phonon scattering in a one-dimensional model with coupling linear in phonon displacements and calculate the probability p(E) of the atom losing energy E as a Fourier transform of a time function $\Pi(\alpha)$. Employing a path integral formulation, and taking the limit of zero de Broglie wavelength of the atom, non-perturbational corrections to the Trajectory Appromixation are obtained as a series in one upon atom mass m. The 1/m terms give rise to α-dependent trajectories, which are shown to be consistent with the TA energy transfer, and lead to energy gain for a slow-particle on a hot surface.

The problem of energy exchange, say in scattering, between atoms or molecules and a surface may be approached at several levels. The most complete is a fully quantum description, which may be essential for light particles, e.g. H_2 or He at thermal energies [1]. At the other extreme all degrees of freedom may be treated classically, [2] as is reasonable for heavy particles, e.g. Xe or A, having strongly multiphonon scattering though it is not appropriate for all observations; one may not calculate the Debye- Waller factor classically, nor any diffraction phenomena. A very useful intermediate level is to treat the particle classically, as is appropriate if it is heavy (Xe,A), but the surface quantally [3]. One may then calculate energy and momentum exchange, and Debye-Waller factors.

In this paper we consider the semiclassical approach just mentioned, in which the particle is treated classically but the surface quantally. The most simple such approach is the trajectory approximation [4,5] in which there is defined for the classical particle a unique trajectory X(t). Then the surface sees a time-dependent perturbation leading to a decoupling of the surface from the particle. For harmonic oscillators in the surface linearly coupled to X(t), this leads to a simple formula which can be very useful in analysing data even in the classical limit [5]. This is demonstrated for rare-gas scattering from transition metal surfaces in Fig. 1, where a considerable amount of data is fitted using one free parameter.

But the very success of the TA raises questions about its validity. What is the equation of motion defining X(t)? If some particles are trapped, how do we cope with a situation with, in reality, two types of trajectory, trapped and reflected? Why does the TA not lead to energy gain on a hot surface?

Evidently a theory going beyond the TA within its classical treatment of the particle could be valuable in answering some of these questions, and we develop one here, described at greater length in ref. 3. We shall simplify the formalism a bit by treating a one-dimensional model, though this restriction is easily lifted.

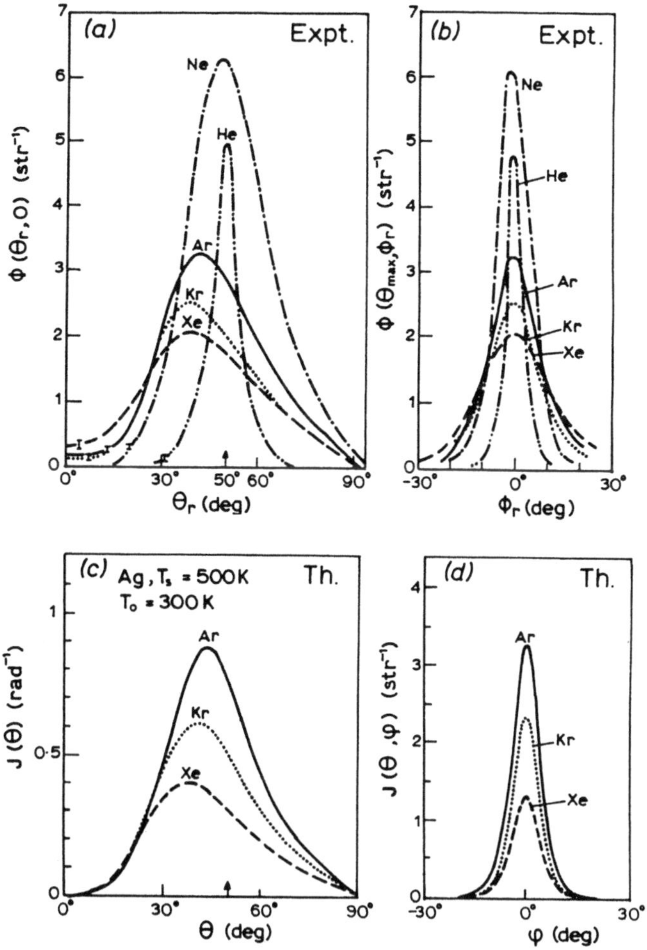

Fig. 1 Comparison of data for rare-gas differential cross-sections on Ag (111) [6] (Θ=angle in scattering plane, Ψ=azimuth), with trajectory approximation calculation of Ref. [5] using $\overline{\Delta E}$ = 215, 415 and 794 °K for Av, Kr, and Xe respectively.

We consider a model, appropriate e.g. for rare-gas atoms at thermal energies on transition metal surfaces, where the surface phonons are coupled linearly to an atomic particle (e.g. a rare gas atom)

$H = H_{osc} + H_p + V$ where

$$H_{osc} = \hbar \sum_k \omega_k \, b_k^+ b_k \tag{1a}$$

$$H_p = \frac{p^2}{2m} + v(x) \tag{1b}$$

$$V = \sum_k \left(\frac{\hbar g}{2\mu\omega_k} \right)^{.5} V_k(x) \, (b_k^+ + b_k) \tag{1c}$$

Here ω_k are the phonon frequencies of a semi-infinite chain of atoms of mass μ modelling the surface, m is particle mass and v(x) the potential seen by the particle when the surface atoms are in their original equilibrium positions, and the V_k's are coupling functions. The b_k's are boson operators for the normal modes and g a coupling constant to be set to unity.

We wish to calculate the probability p(E) of the system gaining energy E from the particle in a collision. p(E) is written as a Fourier transform

$$p(E) = \frac{1}{h} \int_{-\infty}^{\infty} \exp\left[-iE\alpha/h\right] \Pi(\alpha)d\alpha \tag{2}$$

where the "time function" Π is defined as

$$\Pi(\alpha) = Z^{-1}\sum e^{-\beta_m} \langle m,t_f|\exp\left[i\alpha\hbar^{-1}(H_{osc} - \varepsilon_m)\right]|t_f,m\rangle \tag{3}$$

In (3) $\beta=1/kT$, ε_m are eigenstates of H_{osc}, Z the surface p.f., and t_f a time at which the particle has left the surface. $|t_f,m\rangle$ is the state evolving from $|m\rangle$ at the initial time t_i.

In trajectory approximation a unique trajectory X(t) is introduced, and the well known solution for $\Pi(\alpha)$ then is

$$\Pi_{TA}(\alpha) = \exp\{ \frac{-g}{2\mu\hbar} \sum|V_k(\omega_k)|^2\omega_k^{-1}[\mathrm{cth} \frac{\hbar\omega_k}{2kT} (1-\cos\omega_k\alpha) \tag{4}$$

with mean energy-loss

$$\overline{\Delta E} = \int p(E)\, E\, dE = i\hbar\Pi'(0) = \frac{g}{2\mu} \sum_k V_k(\omega_k)^2 \tag{5}$$

and Debye-Waller factor e^{-2W} where

$$2W = \frac{g}{2\mu\hbar} \sum_k V_k(\omega_k)^2 \omega_k^{-1} \mathrm{cth} \frac{\hbar\omega_k}{2kT} \tag{6}$$

Here $V_k(\omega_k)$ is

$$V_k(\omega_k) = \int V_k(X(t))e^{i\omega_k t} dt \tag{7}$$

The time function Π can be rewritten, introducing evolution operators to get $|t_f,m\rangle$ in terms of $|t_i,m\rangle$ and writing the evolution operators in Feynman path integral formalism, as

$$\Pi(\alpha) = \int dx_i \int dy_i\, \Psi_i^*(x_i)\Psi_i(y_i) \int Dx \int Dy$$

$$\langle\exp\{\frac{-i}{\hbar} \int_{t_i}^{t_f} dt[L_p(x) - H_{osc} - V(x)]\}$$

$$\exp\{ \frac{i}{\hbar} \int_{t_i}^{t_f} [L_p(y) - H_{osc} - V_\alpha(y)]\}\rangle \tag{8}$$

Here $\Psi_i(x_i)$ is the initial (incoming) wavefunction of the particle in terms of its initial position variable x_i, $L_p(x) = .5m\dot{x}^2 - v(x)$, and Dx, Dy imply path integration over x(t) and y(t), the paths having a common termination at $x(t_f) = y(t_f)$ but different starting points at $x_i = x(t_i)$, $y_i = y(t_i)$. $V_\alpha(y)$ is the perturbation displaced by time α

$$V_\alpha = \exp[-iH_{osc}\,\alpha/\hbar]\, V \exp[iH_{osc}\,\alpha/\hbar]. \tag{9}$$

The notation < > implies thermal averaging with respect to H_{osc} of the b_k operators in (8).

Next we carry out the thermal average under the path integral, to obtain

$$\Pi(\alpha) = \int dx_i \int dy_i\, \Psi^*(x_i)\Psi(y_i) \int Dx \int Dy \exp[iS/h] \tag{10}$$

where

$S = S_o + S_{int}$, and

$$S_o = -\int_{t_i}^{t_f} [m\dot{x}^2 - m\dot{y}^2 - v(x) + v(y)]dt \tag{11}$$

with

$$S_{int} = ig\int dt \int ds \sum \{\exp[-i\omega_k(t-s)[V_k(x_t)V_k(x_s)\Theta(t-s) + V_k(y_t)V_k(y_s)\Theta(s-t) - \exp(i\omega_k\alpha)V_k(x_t)V_k(y_s)]n_{k-} + [\text{c.c. term}]n_{k+}\}/2\mu\omega_k \tag{12}$$

and

$$n_{k\pm} = \pm[\exp(\pm\hbar\beta\omega_k) - 1]^{-1} \tag{13}$$

(10-13) is formally exact.

It is now convenient to write the result (10-13) in terms of sum and difference coordinates, $q = x-y$, $R = (x+y)/2$. The result is approximated by assuming q small and expanding in powers of q, the implications of which are discussed shortly. At the same time it is convenient to insert the boundary conditions on the paths q(t) and R(t). We assume that the initial wavefunction $\Psi_i(x)$ has the form $\Psi_i(x) \simeq \exp(ik_x x)$, when

$$\Pi(\alpha) = \int DR \int Dq \exp[i\hbar^{-1}(S_o + S_{int})] \tag{14}$$

the path q(t) being unrestricted, while R(t) should obey $m\dot{R}(t_i) = -hk$. Now

$$S_o = \int_{-\infty}^{\infty} dt\; q[m\ddot{R} + v_1(R) + O(q^2)], \tag{15}$$

while

$$g^{-1} S_{int} = \iint q_t\, B(t,s) q_s\, ds dt + \iint C(t,s) ds dt - \iint q_t\, \partial Q(t,s)/\partial R_t\, ds dt + O(q^3) \tag{16}$$

where the response functions B,C,Q are defined by

$$B(t,s) = i(8\mu)^{-1} \sum V_{k1}(R_t) \cos\omega_k(t-s)[(1+\cos\omega_k\alpha)\coth\frac{\hbar\omega_k}{2kT} + i\sin\omega_k\alpha]\omega_k^{-1} V_{k1}(R_s) \tag{17}$$

$$C(t,s) = i(2\mu)^{-1} \sum_k V_k(R_t) \cos\omega_k(t-s)\left[(1-\cos\omega_k\alpha)\, \mathrm{cth}\, \frac{\hbar\omega_k}{2kT} - i\, \sin\omega_k\alpha\right]\omega_k^{-1}\; V_k(R_s) \qquad (18)$$

$$Q(t,s) = (2\mu)^{-1} \sum_k V_k(R_t)\, \sin\omega_k(t-s)\left[(\mathrm{sgn}(t-s)+\cos\omega_k\alpha) + i\, \sin\omega_k\, \alpha\, \mathrm{cth}\frac{\hbar\omega_k}{2kT}\right]\; \omega_k^{-1} V_k(R_s) \qquad (19)$$

In these equations a suffix $_1$ implies differentiation, e.g. $v_1 = \partial v/\partial R$.

The rationale behind our expansion in q is seen by just looking at the S_0 term in (14). If we drop the term of order q^3 in the Lagrangian and carry out the path integral over q, it will give $\prod_t \delta(m\ddot{R}(t) + v_1(R(t)))$ i.e. it will enforce the classical equation of motion on the particle. One may show that including the $O(q^3)$ term leads to quantum corrections to the motion of the particle.

We shall assume that this procedure, which is valid when S_{int} is neglected, is also valid in the presence of S_{int}, i.e. we shall next do the q-path integral in (14). This is analogous to Schmidt's quasi-classical approximation [7]. The result of this procedure is

$$\Pi(\alpha) = \int DR \exp\left\{i\hbar^{-1}\left[-(4g)^{-1}\iint\xi_s\; K(s,t)\xi_t\; ds dt + g\iint C(s,t)\; ds dt + \frac{i}{2} h\; \mathrm{Trln}\; B\right]\right\} \qquad (20)$$

where K is the solution to

$$\int K(s,\tau)\; B(\tau,t)d\tau = \delta(s-t). \qquad (21)$$

An alternative form of (20) is obtained by changing to the variable $\xi(t)$, needing introduction of the Jacobian of the transformation leading to

$$\Pi(\alpha) = \int D\xi \exp\left\{i\hbar^{-1}\left[-(4g)^{-1}\iint\xi_s K_\alpha(s,t)\xi_t ds dt + g\iint C_\alpha(s,t) ds dt\; \frac{i}{2}\hbar\; \mathrm{Trln}B + i\hbar \mathrm{Trln}(I-G^0P)\right]\right. \qquad (22)$$

Here $P(t,s) = \partial^2 Q(t,s)/\partial R_t\, \partial R_s$, and G^0 is the solution to $(v_2=\partial^2 v/\partial R^2)$

$$\left[m\partial^2/\partial t^2 + v_2(R_t)\right]\; G^0(t,s) = \delta(t-s), \qquad (23)$$

subject to

$$G^0(-\infty,s) = dG^0(t,s)/dt\big|_{t=-\infty}=0.$$

We now believe that we have achieved an expression (22), in terms of a path integral, of the problem with the particle's degree of freedom treated classically, provided that the effects of the interaction part S_{int} of the action do not change the meaning of this statement. One could, in principle, stop at this level and do a Monte Carlo treatment of (22), which would, if it could be implemented, be an extension of existing techniques which treat all degrees of freedom classically.

But to give some meaning to (22), let us calculate (22) up to order g^2 in the exponent, which involves doing Gaussian fluctuations around the stationary paths, denoted by $X_\alpha(t)$. To the accuracy specified*, the stationary path obeys the following equation of motion

$$m\ddot{X}_{\alpha t} + \partial\left[v(X_{\alpha t}) - g\int Q_\alpha(T,s)ds\right]/\partial X_{\alpha t} = 0, \tag{24}$$

obtained by taking the stationary point of the action w.r.t. R(t), and ignoring corrections of order g^3. $X_{\alpha t}$ is to be used in calculating Q according to (19). Note that $X_\alpha(t)$ is α-dependent, unlike the unique trajectory in the trajectory approximation, an important point which we return to later. For general α, X_α is also complex.

The corrections from Gaussian fluctuations lead to the following for $\Pi(\alpha)$

$$\Pi(\alpha) = \exp\Big\{ \frac{i}{\hbar} g\iint C_\alpha(s.t)dsdt + g\ \mathrm{Tr}\ G^0P_\alpha + \cdot 5g^2\mathrm{Tr}(G^0P_\alpha G^0P_\alpha)$$
$$+ g^2\mathrm{Tr}(B_\alpha G^0D_\alpha G^0) + O(g^3)\Big\}, \tag{25}$$

Where $D = \partial^2C(t,s)/\partial R_t\partial R_s$. The result (25) can be <u>qualitatively</u> represented by the diagrams in Fig. 2. They describe successive scatterings of the particle described by its classical propagator G^0 by interaction with the surface propagators P, B, D etc. Since G^0 is seen from (23) to be of order 1/m (let us formally classify v to be of order m), it is most helpful to classify the terms of the series by powers of 1/m (m = particle mass).

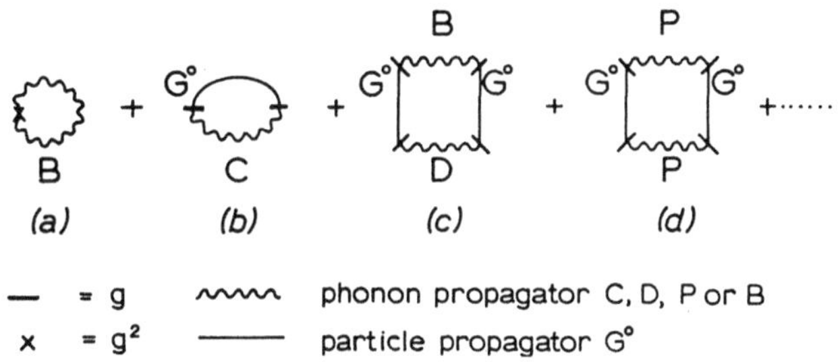

<u>Fig. 2</u> Informal representation of terms in expansion (Eq. 25) of $\ln\Pi(\alpha)$ by diagrams.

The leading power of 1/m involves all but the first term in C in (25) and defining the trajectory (24) just by $m\ddot{X} + v_1(X) = 0$. Inserting the definition (18) of C into the integral in (25), we get back the TA (4).

The corrections in $(1/m)^n$, for $n > 0$ involve various kinds of fluctuations about the unique trajectory X(t). Consider O(1/m). In this order we have the effective potential correction coming into the equation of motion (24) for X_α. We also have the term

$$g\mathrm{Tr}G^0P_\alpha = g\iint G^0(s,t)P_\alpha(t,s)\ dsdt, \tag{26}$$

which seems to represent recoil. Let us look at these terms in the order mentioned.

An obvious test of the effective potential is to calculate the work w_α done against it; if we do this we get two physical limits

---
* Also this stationary point is correct to O(1/m) in the final expression (25).

$$w_0 = \frac{g}{2\mu} \sum_k |V_k(\omega_k)|^2$$

$$w_\infty = 0 \tag{27}$$

The first result is exactly equal to the mean energy-loss (5) of the particle in TA. Since the mean energy-loss indeed corresponds to $\alpha \to 0$ (see (5)), this means that for $\alpha \to 0$ the energy-loss along X is consistent with that calculated from $\Pi(\alpha)$ to the same order in 1/m. To interpret the second result we need to understand the meaning of $\alpha \to \infty$; for continuous distributions of phonon frequency ω_k this is a well-defined limit, and in fact $\Pi(\alpha)$ is the Debye-Waller factor e^{-2W}. Therefore again the work done along X is consistent with that (i.e. zero) expected from $\Pi(\alpha)$ as $\alpha \to \infty$, the Debye-Waller factor being the intensity of the elastic peak E = 0.

Another limit is the Inverse Born-Oppenheimer one where the phonon times ω_k^{-1} are short compared with particle motion time. Taking the large-ω_k limit of Q then (24) becomes

$$m\ddot{X}_{\alpha t} + \partial\left[v(X_{\alpha t}) - \frac{g}{2\mu} \sum_k V_k^2(X_{\alpha t})\omega_k^{-2}\right] = 0 \tag{28}$$

In (28) we get back the straightforward modification in v which can be got by completing the square in the Hamiltonian, neglecting particle kinetic energy.

The Recoil term may be calculated for a "free particle" model, v(x) = 0, when we obtain

$$G^0(t,s) = m^{-1}\Theta(t-s)(t-s) \tag{29}$$

which may be inserted into (26). We use this to calculate the contribution of (26) to the mean energy-loss and obtain the term

$$\overline{\Delta E_R} = \frac{gh}{4m\mu} \sum_k \mathrm{cth}\, \frac{\hbar\omega_k}{2kT} \frac{\partial}{\partial\omega_k} |V_{k_1}(\omega_k)|^2. \tag{30}$$

$$\simeq \frac{gkT}{2m\mu} \sum_k \omega_k^{-1}\partial|V_k(\omega_k)|^2/\partial\omega_k, \quad \text{if } kT \gg h\omega_k. \tag{31}$$

Eq. (31) is especially interesting in the high temperature limit. It leads for sufficiently slow particles to an <u>energy gain</u>, and at large enough T this will dominate the loss (5).

Hence we see that the α-dependent trajectory (24) and the recoil term (26) correct some of the deficiencies in the Trajectory Approximation. The trajectory now becomes more consistent with the results calculated from it using $\Pi(\alpha)$. Also energy-gain appears for a slow particle or a hot surface, a feature abseent in the TA which always gives the <u>loss</u> (5). The higher corrections (third and fourth terms in (25)) represent higher order corrections to th e trajectory. However it seems worth trying to investigate numerically the effects of including just the 1/m corrections, i.e. the self-consistent trajectory (24) and the recoil term (26). We call such an approximation which selects the first two terms in (25) the Improved Trajectory Approximation, and hope to apply it in the future.

I should like to acknowledge the hospitality of the Institute for Theoretical Physics, Santa Barbara, where this work was initiated, and my long-suffering secretary who struggled with this paper.

References

1 V. Celli and D.Evans: Dynamics of Gas-Surface Interactions (Springer-(Springer-Verlag 1982).

2 J.C. Tully: J. Chem. Phys. 73, 1975 (1980)

3 D.M. Newns: submitted to Phys. Rev

4 W. Brenig: Z. Physik B36, 81 (1979); R. Sedlmeir and W. Brenig: Z. Physik B36, 245 (1980); J. Boheim and W. Brenig: Z. Physik B41, 243 (1981)

5 R. Brako and D.M. Newns: Phys. Rev. Lett. 48, 1859 (1982); R Brako and D.M.Newns: Surface Sci. 117, 42 (1982); R. Brako: Surface Sci. 123, 439 (1982)

6 H. Asada: Japan J. Appl. Phys. 19, 2055 (1980)

7 A. Schmidt: J. Low Temp. Phys. 49 608 (1982)

Theory of Surface Dynamical Processes with Gas-Solid Electron Transfer

M. Tsukada and N. Shima

Department of Physics, University of Tokyo
Bunkyo-ku, Tokyo 113, Japan

1. Introduction

Many dynamical phenomena on surfaces involve electron transfer between particle and surface as a crucial elementary process. For instance, field desorption [1], gas/surface charge exchange scattering [2], chemisorption of oxygen or halogen on metals [3], deexcitation process of ESD(PSD) [4] belong to such kind of phenomena, as well as electrode processes in electro-chemical cells. These phenomena are essentially non-adiabatic, since the time scale of the electron transfer is comparable or longer as compared to that of the adsorbate motion.

The simple models, which have been studied most intensively, treat the adsorbate motion in a semiclassical way and electrons with the independent particle approximation. Although there have been considerable efforts to clarify the electron transfer mechanism in such models, they are directed only to certain limiting cases, i.e. the wide-band-limit [2] and the limit of the discrete level crossover [5]. The relation between the two limits is not yet established. In the first part of the paper, we will study the electron transfer process for general cases by means of the time-developing operator method, and propose a simple expression of the electron transfer probability. The expression connects the known results of the both limits.

The coupling of electronic states with medium, such as phonons, plasmons, electron-hole excitations and solvate configurations significantly influences the electron transfer process. This effect is studied in the present paper by the time-dependent-Hartree method [6] and by the time-developing operator method. The important finding is that the electron transfer process is drastically suppressed, if the coupling with the medium is strong enough.

For the case of field desorption, the so-called adsorbate 'interband tunneling' accompanied by the electron transfer is important. Such phenomena cannot be treated within the framework of the semiclassical theory and one must use full quantum mechanical approach. The detailed discussion for this problem is described in the next paper [1].

2. Electron Tunneling Probability by the Semiclassical Independent Particle Theory

2.1 Model of the Electron Transfer

In the semiclassical independent particle model, the electron transfer is described by the following Hamiltonian,

$$H_e(t)=H_o(t)+V(t) \quad , \tag{2.1}$$

$$H_o(t)=\varepsilon_o(z(t))a_o^\dagger a_o+\sum_k \varepsilon_k a_k^\dagger a_k \quad , \tag{2.2}$$

$$V(t)=\sum_k V_k(z(t))a_o^\dagger a_k + h.c. \quad . \tag{2.3}$$

Relevant quantities for the electron transfer process are the expectation values of the operators for the initial state,

$$n_{ij}(t)=U_e^\dagger(t,t_o)a_i^\dagger a_j U_e(t,t_o) \qquad (i,j=0,k_1,k_2,\ldots) \quad , \tag{2.4}$$

where

$$U_e(t,t_o)=T_+\exp(-\frac{i}{\hbar}\int_{t_o}^{t} H_e(\tau)d\tau) \quad . \tag{2.5}$$

Equation of motion of $n_{ij}^{I}(t)$ defined by

$$n_{ij}^{I}(t)=n_{ij}(t)\exp(-\frac{i}{\hbar}\int_{t_o}^{t} \varepsilon_{ij}(\tau)d\tau) \quad , \tag{2.6}$$

$$\varepsilon_{ij}=\varepsilon_i-\varepsilon_j \quad , \tag{2.7}$$

is written as

$$i\hbar\dot{n}_{ij}^{I}=\sum_l (V_{jl}(t)n_{il}^{I}-V_{li}(t)n_{lj}^{I}) \quad , \tag{2.8}$$

where

$$V_{ij}=V_{ji}^{*}=\begin{cases} V_k(z(t))\exp(\frac{i}{\hbar}\int_{t_o}^{t} \varepsilon_{ok}(\tau)d\tau) & (i=0,j=k) \\ 0 & (\text{otherwise}) \quad . \end{cases} \tag{2.9}$$

Equation (2.8) is integrated in the form of

$$n_{ij}^{I}(t)=\sum_{lm} U_{il}(t,t_o)n_{lm}^{I}(t_o)U_{mj}^{\dagger}(t,t_o) \quad , \tag{2.10}$$

with

$$U(t,t_o)=T_+\exp(-\frac{i}{\hbar}\int_{t_o}^{t} V^{I}(t)dt), \tag{2.11}$$

where the matrix $V^I(t)$ is defined by its element V_{ij} of (2.9). Utilizing the special form of $V^I(t)$, we can obtain the elements of the time-developing operator $U(t,t_0)$ as follows:

$$U_{oo}(t,t_o)=1+\sum_{n=1}^{\infty}(\frac{-1}{\hbar^2})^n\int_{t_o}^{t}dt_1 B(t,t_1)\int_{t_o}^{t_1}dt_2 B(t_1,t_2)\cdots\cdot\int_{t_o}^{t_{n-1}}dt_n B(t_{n-1},t_n), \tag{2.12}$$

$$B(t_1,t_2)=\int_{t_2}^{t_1}d\tau\sum_k V_{ok}(\tau)V_{ko}(t_2) \quad , \tag{2.13}$$

$$U_{ok}(t,t_o)=-\frac{i}{\hbar}\int_{t_o}^{t} d\tau U_{oo}(t,\tau)V_{ok}(\tau) \quad , \tag{2.14}$$

$$U_{kk'}(t,t_o)=\delta_{kk'}-\frac{1}{\hbar^2}\int_{t_o}^{t}d\tau\int_{t_o}^{\tau}d\tau' V_{ko}(\tau)U_{oo}(\tau,\tau')V_{ok'}(\tau') \ . \tag{2.15}$$

The occupation numbers of the state 0, k, are easily obtained by

$$n_o(t)=n_o^o|U_{oo}(t,t_o)|^2+\sum_k n_k^o\,|U_{ok}(t,t_o)|^2 \ , \tag{2.16}$$

$$n_k(t)=n_o^o|U_{ko}(t,t_o)|^2+\sum_{k'} n_{k'}^o|U_{kk'}(t,t_o)|^2 \ , \tag{2.17}$$

where n_0^0, n_k^0 are the initial values of n_0, n_k.

2.2 An Approximation to $U_{00}(t,t_0)$

A practical and efficient approximation to sum up the infinite series of U_{00} is to substitute

$$B_o(t_2;t^*)=\int_{t_2}^{t^*}d\tau\sum_k V_{ok}(\tau)V_{ko}(t_2) \ , \tag{2.18}$$

for $B(t_1,t_2)$ in (2.12), where t^* is fixed at the time very close to t. For $t_2<t_1<t^*$, the difference $B(t_1,t_2)-B_0(t_2;t^*)$ is not significant as compared to $B_0(t_2;t^*)$, since the function $\sum_k V_{0k}(\tau)V_{k0}(t)$ rapidly oscillates with the increase of $|\tau-t|$. If necessary, corrections caused by $B(t_1,t_2)-B_0(t_2,t^*)$ can be estimated. After taking the limit $t^*\to t$, we obtain

$$U_{oo}(t,t_o)=\exp[-\frac{1}{\hbar^2}\int_{t_o}^{t}dt_1\int_{t_o}^{t_1}dt_2\sum_k V_k(t_1)V_k^*(t_2)\exp(\frac{i}{\hbar}\int_{t_2}^{t_1}\varepsilon_{ok}(\tau)d\tau)] \ . \tag{2.19}$$

2.3 Crossing of the Occupied Discrete Level with an Empty Band

For a simple example, consider the case

$$\begin{cases}\varepsilon_a(z(t))=Fz(t)=Fvt\\ V_k(z(t))=V_k\exp(-az^2(t))\\ \Delta(\omega)=\pi\sum_k|V_k|^2\delta(\omega-\varepsilon_k)=\pi|V|^2/\Delta\sqrt{\pi}e^{-\omega^2/\Delta^2}\end{cases} \ . \tag{2.20}$$

Then the above-mentioned approximation results in

$$U_{oo}(\infty,-\infty)=\exp[-\frac{2\pi|V|^2}{\hbar v}/\sqrt{4\hbar^2v^2a^2+2a\Delta^2+F^2}] \ . \tag{2.21}$$

For the case of the time-independent coupling constant (a=0), the above is just the LANDAU-ZENER result [5]. It is remarkable that the result does not depend on the width of the band. For the case $\Delta=0$, $a\neq0$, (2.21) is also obtained rigorously by the Zener's method.

The wide band approximation, slightly generalized as the original one proposed by BRAKO and NEWNS [2] leads to

$$n_o(\infty)=n_o(-\infty)\exp(-\frac{2\pi|V|^2}{\hbar v}/\sqrt{2a\Delta^2+F^2}) \ . \qquad (2.22)$$

As compared to (2.21), the term of $4\hbar^2v^2a^2$ is dropped from the root of the denominator of the exponent. Thus the wide band limit fails, unless the band width is larger than $\hbar/\tau_{cross}$ ($\tau_{cross}=\sqrt{1/a}/v$; the time needed for the crossover).

2.4 General Expression for the Charge Transfer Probability

Relation (2.19) is rewritten as

$$U_{oo}(t,t_o)=\prod_k U_{oo}(t,t_o;k) \ , \qquad (2.23)$$

$$U_{oo}(t,t_o;k)\equiv\exp[-\frac{1}{\hbar^2}\int_{t_o}^{t}dt_1\int_{t_o}^{t_1}dt_2V_k(t_1)V_k^*(t_2)e^{-\frac{i}{\hbar}\int_{t_2}^{t_1}\varepsilon_{ko}(\tau)d\tau}] \ . \qquad (2.24)$$

This allows the simple interpretation that the discrete-continuum crossover is a cascade of pairwise level crossing. For the linear crossover as shown in Fig.1,

$$|U_{oo}(t,t_o;k)|^2=\exp[-\frac{1}{\hbar^2v^2}|\int_{z(t_o)}^{z(t)}duV(u)e^{-i(\Delta/2\hbar v)u^2}|^2] \ , \qquad (2.25)$$

where the origin of the reaction coordinate z(t) is chosen as the crossover point z_k. The above U_{OO} results in the exact Landau-Zener result in the long time limit ($t\to\infty$) and for the constant coupling, V(u)=const. From the relations (2.23-25), the simple semiclassical result

$$n_o(t=\infty)=n_o^o\exp(-\int_{z_o}^{\infty}\xi(z)dz)+\int_{z_o}^{\infty}f(\varepsilon_o(u))\xi(u)\exp(-\int_u^{\infty}\xi(v)dv)du \ , \qquad (2.26)$$

is obtained, where $z_O=z(t_O)$, f is the Fermi distribution function and

$$\xi(z)=\frac{2\pi}{\hbar v}\sum_k|V_k(z)|^2\alpha_k\delta(\varepsilon_k-\varepsilon_o(z)) \ , \qquad (2.27)$$

$$\alpha_k=\frac{\Delta}{2\pi\hbar v}|\int_{-\infty}^{\infty}du(V_k(z_k+u)/V_k(z_k))\exp(-i\frac{\Delta}{2\hbar v}u^2)|^2 \ . \qquad (2.28)$$

For exponentially decaying coupling ($V_k(z)\sim e^{-cz}$), α_k is equal to unity. Equations (2.26-28) generalize the expression of the wide-band limit [2] for $T\gg T_{eff}=\hbar\frac{d}{dz}\ln\varepsilon_o(z)/\pi k_Bv$, for the case of the general bands. These relations are also the generalization of the Landau-Zener theory [5], since they reproduce the latter in the narrow band limit. It should be noted that the relation (2.26) can be obtained from the solution of

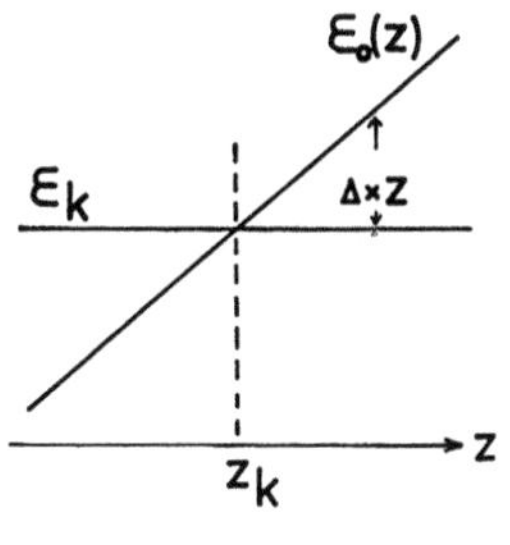

Fig.1 Crossover of the adsorbate valence level $\varepsilon_O(z)$ with a k-th band state

$$\frac{dn_o(z)}{dz} = -\xi(z)n_o(z)+f(\varepsilon_o(z))\xi(z) \ , \tag{2.29}$$

by taking the limit of $z\to\infty$. As seen in (2.27), $\xi(z)$ is similar to, but not the same as, the transition probability of the golden-rule divided by velocity.

3. Effect of the Coupling with Medium

In the real cases, the adsorbate system couples with various modes of the medium, such as phonons, surface plasmons, electron-hole pair excitations. The coupling with medium significantly influences the charge-transfer process. We analyse such effects by Hamiltonian

$$H_{eM}=H_e+\sum_\lambda \hbar\omega_\lambda b_\lambda^\dagger b_\lambda+\sum_\lambda \eta_\lambda(z(t))(b_\lambda^\dagger+b_\lambda)a_o^\dagger a_o \ , \tag{3.1}$$

where H_e is given by (2.1) and the medium modes are treated by the Boson model coupled linearly with adsorbate charge.

3.1 The Static and Narrow Band Limit

In the static case ($z(t)$=const) and at the narrow band limit ($\varepsilon_k=\varepsilon_1$=const), H_e reduces to the Hamiltonian of the simple molecule

$$H_e=\varepsilon_o a_o^\dagger a_o+\varepsilon_1 a_1^\dagger a_1+V(a_o^\dagger a_1+a_1^\dagger a_o) \ . \tag{3.2}$$

If the coupling with the medium is strong, the molecular orbital is not formed due to the hindered electron transfer between the two sites. To see this we use the variational wave function

$$\Psi(\alpha,u) = \cos\alpha \ \phi_o \prod_\lambda e^{\gamma_\lambda u(b_\lambda-b_\lambda^\dagger)}|0\rangle + \sin\alpha \ \phi_1|0\rangle \ , \tag{3.3}$$

where α, u are the variational parameters, $\gamma_\lambda=\eta_\lambda/\hbar\omega_\lambda$ and $|0\rangle$ is the Boson ground state. For the case of $\varepsilon_0=\varepsilon_1$, the energy corresponding to $\Psi(\alpha,u)$ takes a minimum ($\sim\varepsilon_0-|V|$) at (α,u) $(\pi/4,0)$ for $|V|>\bar{\omega}$. On the other hand it takes a minimum ($\sim\varepsilon_0-\bar{\omega}$) at $(\alpha,u)\sim(0,1)$ for $|V|<\bar{\omega}$. In the above $\bar{\omega}$ is the polaron formation energy defined by $\bar{\omega}=\sum_\lambda\hbar\omega_\lambda\gamma_\lambda^2$. In the latter case the electron exchange between the two sites is prohibited and the molecular orbital state is not formed.

3.2 TDH (Time-Dependent-Hartree) Approximation

If the coupling with the heat bath is weak, the so called TDH approximation is effective. In this method the coupling term is approximated as
$(b_\lambda^\dagger+b_\lambda)a_o^\dagger a_o \sim\langle b_\lambda^\dagger+b_\lambda\rangle a_o^\dagger a_o+\langle a_o^\dagger a_o\rangle(b_\lambda^\dagger+b_\lambda)$.
The equation of motion of the expansion coefficients of the electronic state is numerically integrated by the stochastic trajectory approach. Figure 2 shows the example of the obtained survival probability for the case of the discrete two-level crossover. The results of the numerical calculations [6] as well as an analytic considerations indicate that the effective adiabatic parameter ξ^*,

$$\xi^*=\xi/(1-(\bar{\omega}/\Delta)\frac{d}{dz}n_o(z_c)) \ , \tag{3.4}$$

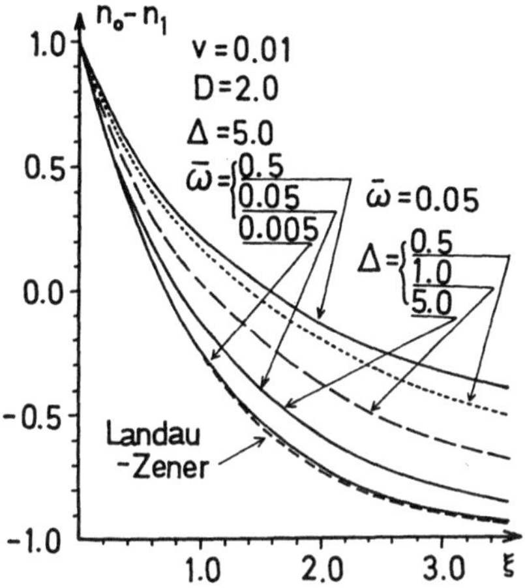

Fig.2 The TDH numerical results of the occupation probability n_0-n_1 after the linear crossover. The initial condition is $n_0(t_0)=1$, $n_1(t_0)=0$, where the state 0 is the lowest state before crossover. The width of the boson spectrum D is assumed to be 2.0 a.u. All the parameters are expressed in a.u. (See [6])

governs the charge transfer probability, where $\xi = 2\pi|V|^2/v\Delta$ is the nominal Landau-Zener adiabatic parameter, $\bar{\omega} = \sum_\lambda \hbar\omega_\lambda \gamma_\lambda^2$ and Δ is the energy gradient at the crossover point z_c. Thus for the case of the crossover from the ground state ($dn_0/dz<0$), the adiabatic passage is significantly enhanced due to the coupling with medium.

3.3 Purturbation Approach

A different approach to this problem which is valid for general cases uses the time-developing operator of section 2. The state vector is expanded as

$$\Psi(t)=C_{oo}(t)|00\rangle+\sum_{k\{n_\lambda\}} C_{k\{n_\lambda\}}(t)|k\{n_\lambda\}\rangle \ , \tag{3.5}$$

where

$$|00\rangle=\phi_o \prod_\lambda e^{\gamma_\lambda(b_\lambda-b_\lambda^\dagger)}|0\rangle \ , \ |k\{n_\lambda\}\rangle=\phi_k \prod_\lambda \frac{(b_\lambda^\dagger)^{n_\lambda}}{\sqrt{n_\lambda!}}|0\rangle \ . \tag{3.6}$$

The coefficient vector

$$^t\mathbf{C}(t)\equiv\left(\exp\left(\frac{i}{\hbar}\int_{t_o}^{t} \varepsilon_o(\tau)d\tau\right)C_{oo}(t),\ \exp\left(\frac{i}{\hbar}\int_{t_o}^{t} (\varepsilon_k+\sum_\lambda n_\lambda\hbar\omega_\lambda)d\tau\right)C_{k\{n_\lambda\}}(t)\right) \ , \tag{3.7}$$

is generated by the time-developing matrix $U(t,t_0)$ as $\mathbf{C}(t)=U(t,t_0)\mathbf{C}(t_0)$. The matrix U is given by the same equation as (2.11), if the matrix elements of $V^I(t)$ is defined by

$$V_{ij}=V_{ji}^*=\begin{cases} V_k(z(t))F(\{n_\lambda\})\exp\left(\frac{i}{\hbar}\int_{t_o}^{t}(\varepsilon_o(\tau)-\varepsilon_k-\sum_\lambda n_\lambda\hbar\omega_\lambda)d\tau\right) \\ \qquad \text{for} \quad i=|00\rangle,\ j=|k\{n_\lambda\}\rangle \\ 0 \qquad \text{otherwise} \ . \end{cases} \tag{3.8}$$

In the above,

$$F(\{n_\lambda\})\equiv\langle 0|\prod_\lambda \frac{b_\lambda^{n_\lambda}}{\sqrt{n_\lambda!}} e^{\gamma_\lambda(b_\lambda-b_\lambda^\dagger)}|0\rangle = \prod_\lambda \frac{(-\gamma_\lambda)^{n_\lambda}}{\sqrt{n_\lambda!}} e^{-\gamma_\lambda^2/2} \ , \tag{3.9}$$

represents the overlap of the Boson state with the different center. At the initial state, the electron is assumed to exist on the adatom state ϕ_0 ($C_{00}(t_0)=1$). After the crossover with an empty band, the electron occupation probability on the ϕ_0 state is given by (2.26) with $\xi(z)$ defined by

$$\xi(z)=\frac{2\pi}{\hbar v}\sum_{k\{n_\lambda\}}|V_k(z)|^2\alpha_k|F(\{n_\lambda\})|^2\delta(\varepsilon_o(z)-\varepsilon_k-\sum_\lambda n_\lambda\hbar\omega_\lambda) . \tag{3.10}$$

As clear from (3.10), the result is equivalent to the cascade crossover of the $|00\rangle$ state with all the $|k\,\{n_\lambda\}\rangle$ state. By the use of the Boson spectral function

$$I(\omega)=\frac{1}{2\pi}\int_{-\infty}^{\infty}dt\ e^{-i\omega t}\exp(\sum_\lambda\gamma^2(e^{i\hbar\omega_\lambda t}-1)) , \tag{3.11}$$

$\xi(z)$ is expressed as

$$\xi(z)=\frac{2\pi}{\hbar v}\sum_{k>k_F}|V_k(z)|^2\alpha_k I(\varepsilon_o(z)-\varepsilon_k) . \tag{3.12}$$

For the case of strong coupling with medium, the spectral function is approximately expressed as

$$I(\omega,z)\simeq\frac{1}{\sqrt{\pi}\Gamma(z)}\exp(-(\omega-\bar{\omega}(z))^2/\Gamma^2(z)) , \tag{3.13}$$

where

$$\bar{\omega}(z)=\sum_\lambda\hbar\omega_\lambda\gamma_\lambda^2 , \tag{3.14}$$

$$\Gamma(z)=\sqrt{2\sum_\lambda\eta_\lambda^2(z)} . \tag{3.15}$$

The sum rule of the transition factors,

$$\sum_{\{n_\lambda\}}|F(\{n_\lambda\})|^2=\int I(\omega,z)d\omega=1 , \tag{3.16}$$

should be remarked. The probability of the "zero-Boson" crossing is in proportion to

$$P_o\sim|F(\{n_\lambda\})|^2_{\text{all},\eta_\lambda=o}=\exp(-\sum_\lambda\frac{1}{2}\frac{\eta_\lambda^2}{(\hbar\omega_\lambda)^2}) . \tag{3.17}$$

If the medium is the particle-hole excitation systems modelled by the Tomonaga Bosons, P_0 vanishes because of the linear behavior ($\rho(\omega_\lambda)\sim\omega_\lambda/D$) of the Boson spectrum. This is consistent with the Anderson's orthogonalization theorem.

The physical implication of the result is illustrated in Fig.3. Without the coupling with the medium, the crossover of ϕ_0 with a substrate state ϕ_k takes place at $z=z_0$. By the effect of the coupling, however, the crossover points are distributed continuously in the region $z>z_0$. Considering the sum rule of $|F(\{n_\lambda\})|^2$ and the rapid decrease of the transfer integral $|V_k(z)|^2$ with z, one may conclude that the coupling to the medium significantly suppresses the electron transfer probability between the adatom and the substrate. This conclusion is qualitatively the same as that obtained by TDH [6].

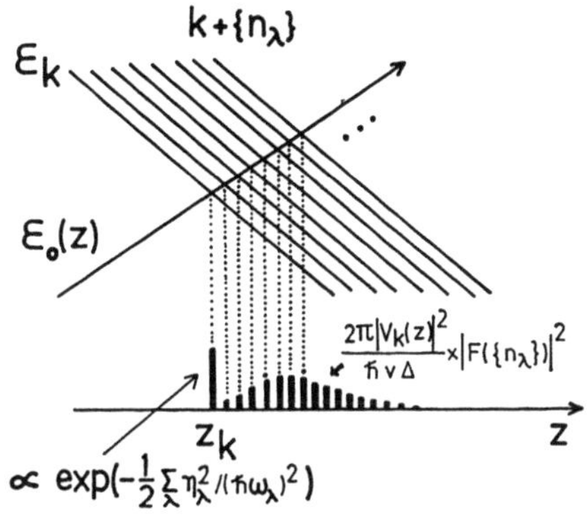

Fig.3 Effect of the boson coupling on the crossover of the adsorbate state $\phi_0(z)$ with a continuum state ϕ_k

3.4 Application to the Field Desorption Process

In a simple model of the field desorption, the energy levels near the critical crossover region are assumed as shown in Fig.4. Relevant quantities are chosen as follows:

$$\begin{cases} \varepsilon_o(z)-\varepsilon_k=\varepsilon_o-\varepsilon_k+eFz \\ V_k(z)=V_k^o\exp(-sz/2) \\ \eta_\lambda(z)=\eta_\lambda^o\exp(-\gamma z/2) \\ |V|^2\rho\equiv\sum_k|V_k^o|^2\delta(\omega-\varepsilon_k) \; . \end{cases} \tag{3.18}$$

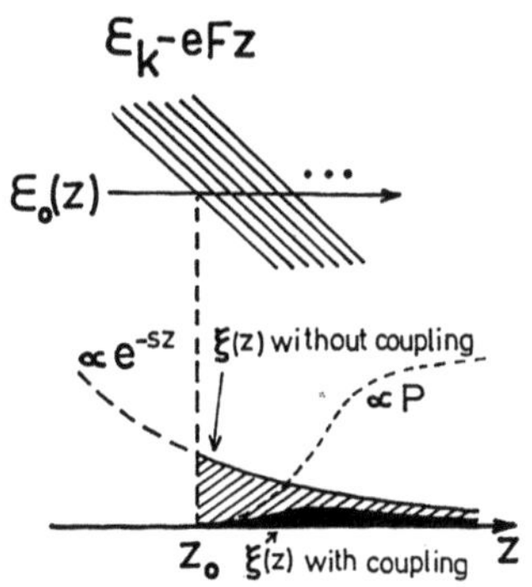

Fig.4 Effect of the boson coupling on (z) determining the transition rate of the field desorption

It is easily obtained that

$$\xi(z)=(2\pi|V|^2\rho/\hbar v)e^{-sz}\;P\left(\frac{\varepsilon_F+\omega e^{-\gamma z}-\varepsilon_o-eFz}{\sqrt{2}\Gamma_o e^{-\gamma z/2}}\right) , \tag{3.19}$$

with

$$P(x)=\frac{1}{\sqrt{2\pi}}\int_x^\infty \exp(-t^2/2)dt \; . \tag{3.20}$$

The probability that the adsorbate remains neutral at $z\to\infty$ is given by

$$n_o(\infty)=\exp\left(-\int_{z_o}^\infty \xi(z)dz\right) \; . \tag{3.21}$$

Figure 4 shows that the above quantity is drastically reduced, if the coupling with the medium is significant. Such effect qualitatively explains the reason why the weak chemisorption bond is hard to be fragmented in the field desorption process, while the strong chemisorption bond is easily dissociated [7].

4. Discussions and Conclusions

In the previous sections the time development of the 'density operator' $a_i^+a_j$ were obtained for the semiclassical independent particle model by the approximate summation of the infinite series of the U matrix. The obtained gas-surface electron transfer probability indicates that the crossing of the discrete level with a continuous band is equivalent to the cascade two-level crossover. The expression of the probability takes a simple form which is similar to but not the same as that obtained by the phenomenological approach based on the Fermi's golden rule. This expression coincides with the result by the wide-band limit in certain cases and the Landau-Zener result for the narrow band limit. The obtained result might be effectively applied to the charge transfer processes on semiconductor surfaces.

For the discussions of the energy-loss spectrum, we can utilize the products of the density operators. Following a similar procedures made in section 2, the loss spectrum can be obtained by the direct product of the U matrix. Such discussions will be made elsewhere.

Coupling with heat-bath or medium modes is found to suppress the charge-transfer process remarkably. If the transfer integral between the adatom and the surface is smaller than the polaron formation energy, the covalent orbital mixing between the two is not realized. Such might be the case if the adsorbate located distant from the surface. By the use of the TDH method and also by the time-developing operator method, the suppression of the electron-transfer by the coupling to the medium is demonstrated. Such effect should be taken into account in considering the realistic model of the non-adiabatic surface processes.

References

1 N.Shima, M.Tsukada: "Dynamical Processes and Ordering on Solid Surfaces", (Springer, 1984)
2 R.Brako, D.M.Newns: Surf. Sci. 108, 253 (1981)
3 J.N.Nørskov, D.M.Newns, B.I.Lundqvist: Surf. Sci 80, 179 (1979)
4 W.Brenig: Z. Phys. B23, 361 (1976)
5 L.D.Landau, E.M.Lifshitz: "Quantm Mechanics", (Addison-Wesley, Reading 1958)
C.Zener: Proc. Roy. Soc. London A137, 696 (1932)
6 M.Tsukada: J. Phys. Soc. Jpn. 51, 2927 (1982)
7 O.Nishikawa, T.Yoshimura, M.Shibata: Surf.Sci. 124, 440 (1983)

Non-Adiabatic Processes in Field Desorption Phenomena

N. Shima and M. Tsukada

Department of Physics, University of Tokyo
Bunkyo-ku, Tokyo 113, Japan

1. Introduction

The field-induced desorption phenomenon [1] of surface atoms or adsorbed molecules gives direct information about the microscopic states of surface atoms. Recently qualitatively different behaviors were found between strongly adsorbed (chemisorbed) and weakly adsorbed (physisorbed) hydrogen-metal surface systems when the high electric field was applied. In the strongly adsorbed systems of hydrogen on Fe or Ni, H^+ ions were detected [2]. On the other hand, in the weakly adsorbed systems of hydrogen on Al or Cu, dominant desorbed species are metal hydride ions and the proportion of the desorbed H^+ is small. Such behavior is puzzling from the usual view points of the chemical reaction, since the weaker bond might be considered harder to dissociate.

In order to clarify the mechanism and rate of the field desorption, we use a simple diatom model which mimics the real adatom-surface system. Desorption process can be described by the one-dimensional model, if band structure effects for lateral motion are not important as in the present case [4]. In the absence of the field, as shown in Fig.1 a), the ionic branch of the potential energy surface, which is the excited state of the molecule, stays at high-energy position. When the high electric field is applied, as shown in Fig.1 b), the ionic branch crosses the neutral branch at R_c and the ionic state becomes the most stable for the region distant from the surface. Because of the interaction between the ionic and neutral states, the lowest adiabatic potential surface has a peak around R_c. The energy difference between the ionic and the neutral branch of the weakly adsorbed system is much larger than that of the strongly adsorbed system.

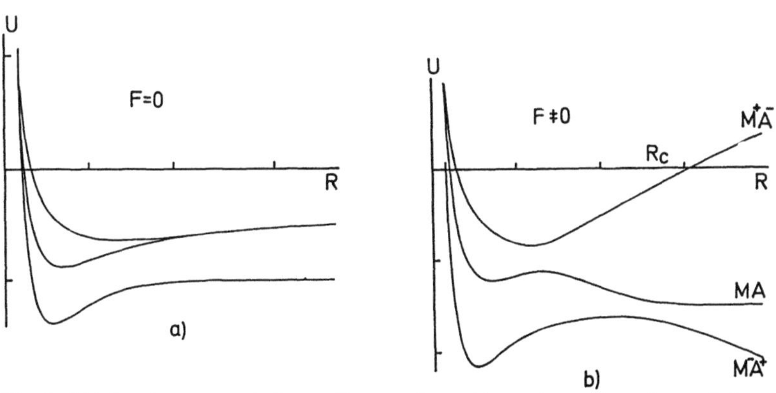

Fig.1 Adiabatic potential surfaces of a diatomic molecule model under the electric field F. a) F=0, b) F≠0

Thus the barrier takes place at relatively distant region from the surface. For the strongly adsorbed case, the position of the peak of the adiabatic potential surface is located at much closer to the substrate surface and the magnitude of the avoid crossing is considerably large. This is because the ionic configurations are significantly admixed in the ground state at the equilibrium distance, and the effect of the field appears already from the region close to the surface.

The desorption processes have been analyzed as a thermally activated process over the potential surface barrier. The desorption rate constant k per an adsorbate is assumed to have a form of $k=\nu\exp(-Q/k_BT)$ where Q is a barrier height depending on the field strength, T is temperature, k_B is Boltzmann constant, and ν is a prefactor depending on the field strength and T [1].

In the case of a strongly adsorbed system, the process follows along the adiabatic potential surface. But in the case of a weakly adsorbed system, the process involves electron tunneling and proceeds not necessarily along the adiabatic potential surface. From the standpoint of this picture, the field desorption is not a simple activation process over the definite potential surface barrier. Rigorously speaking the desorption rate can not be expressed by the simple Arrhenius relation as will be seen in the next section. Thus, if one expresses the desorption rate formally by the Arrhrenius relation, the activation energy Q and the prefactor n depends considerably on T. Moreover, it is possible that the field desorption rate of a weakly (physical) bonding molecule is smaller than that of a strongly (chemical) bonding molecule because of the non-adiabaticity. If this is the case for the hydrogen-metal systems, the field desorption behaviors of the systems are comprehensible.

Theoretically a charge-exchange model of the field evaporation was proposed by GOMER and SWANSON [5] in 1963 considering some part of the above points. But according to their theory, the activation barrier Q which is the difference between the crossing energy and the minimum of the ground adiabatic potential, is fixed. Therefore their theory cannot connect between the adiabatic and diabatic limit correctly. Their expression of the probability of the electronic transition contains nuclear velocity v which is a semiclassical quantity. This is not rigorous for the process around the crossing region. Furthermore the calculated rate constant predict much larger value than the experimental one. These deficiencies might result from their semiclassical perturbative calculation. We investigate here the fundamental problems in field desorption related with such points by quantum non-perturbative calculation paying attention to the effect of the non-adiabaticity.

2. Kinetics of the System

In this section we study the kinetics of our diatomic molecule model system. As shown in Fig.2, the electronic structure of the system under the electric field is simplified as constructed by two electronic states, a neutral state (state 1) and an ionic state (state 2). For the state 1 there are some nuclear vibrational discrete levels $\{\varepsilon_i\}$ (i=0,1,2,..N) and continuum states which are replaced by a single level ε_c here. Experimentally the population number of the state c is set to be constant n_c by supplying neutral gas whose pressure is constant. Desorbing ions are gathered by a negative electrode. We set up master equations between the population numbers of the i-th level, $n_i(t)$. At low coverage of adsorbates, we can

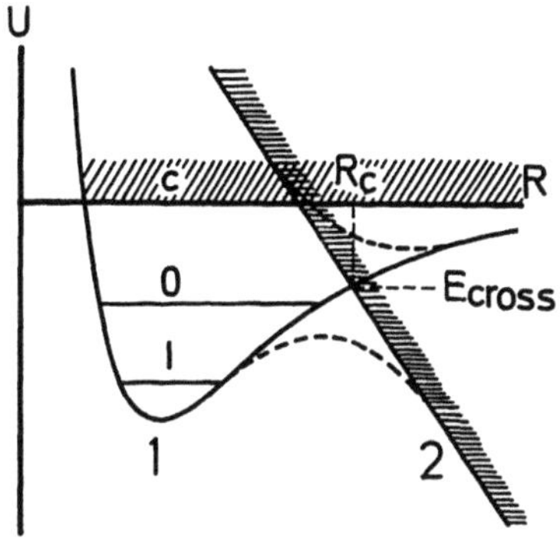

Fig.2 Diabatic potential surfaces (solid line) and vibrational levels of the two electronic states model under the electric field. Hatched area stands for continuum states. Dotted line stands for the adiabatic potential

assume that transitions among the discrete levels are induced only by phonons of the metal surface. Many body excitations of the metal surface are neglected. Then the time-dependence of $n_i(t)$ is governed by the equations,

$$\frac{d}{dt}n_i(t)=\sum_{j\neq i}^{N} R_{ij}n_j(t)+R_{ic}n_c-\sum_{j\neq i}^{N}(R_{ji}+R_{ci})n_i(t)-k_i^{(1)}(t) \qquad (i=0,1,..N), \tag{1}$$

$$k_i^{(1)}(t)=\int_0^t dt'P_i(t-t')\Big[\sum_{j\neq i} R_{ij}n_j(t')+R_{ic}n_c-(\sum_{j\neq i} R_{ji}+R_{ci})n_i(t')\Big]+P_i(t)n_i(0) \ , \tag{2}$$

where R_{ij} is the transition probability by phonon per unit time from the j-th state to the i-th state, and $P_i(t)$ is the decaying rate of the i-th state to the ionic desorbing state after the generation of the i-th state at t=0. Obviously $\{R_{ij}\}$satisfy the equation,

$$R_{ij}=R_{ji}e^{-\beta(\varepsilon_i-\varepsilon_j)}, \quad \beta=\frac{1}{k_BT} \ . \tag{3}$$

As we are interested in the stationary state, t is set to be +∞. With the aid of the final value theorem of Laplace transformation,

$$\sum_{j=0}^{N} D_{ij}n_j(+\infty)=R_{ic}n_c \ , \tag{4}$$

where

$$D_{ij}=\begin{cases}\dfrac{1}{\mu_i}+\sum\limits_{k\neq i}^{N}(R_{ki}+R_{ck}) & i=j\\ -R_{ij} & i\neq j\end{cases} \ , \tag{5}$$

μ_i is the mean first passage time of the i-th level defined by

$$\mu_i=\int_0^\infty dt \ tP_i(t) \ . \tag{6}$$

If all μ_i are set to be +∞, $n_i(+\infty)$ become ordinary Boltzmann distribution. The rate k of detection of ions per an adsorbate is given by

$$k=\frac{\sum_i(k_i^{(1)}(+\infty)+k_i^{(2)}(+\infty))+k_c}{\sum_i n_i(+\infty)} \ , \tag{7}$$

where

$$k_i^{(1)}(+\infty)=\frac{1}{\mu_i}n_i(+\infty)\ , \tag{8}$$

$$k_i^{(2)}(+\infty)=pR_{ci}n_i(+\infty)\ , \tag{9}$$

$$k_c\quad=\alpha pn_c\ , \tag{10}$$

$k_i^{(1)}$ represents the direct ionization fraction from the i-th level and $k_i^{(2)}$ the indirect ionization fraction via the continuum state c from the i-th level. k_c represents the direct ionization fraction from the continuum state. p is an ionization probability of the continuum state and α is a constant. $\{R_{ij}\}$ are given from the phonon structure of the surface metal. $\{\mu_i\}$ are the important quantities to determine the ionic desorption rate k and reflect the microscopic non-adiabatic process which will be discussed in section 3.

To see a qualitative feature we apply the above expression to the simple case where N=1 ($\varepsilon_c>\varepsilon_0>\varepsilon_1$) and $\mu_0=\mu$, $\mu_1=+\infty$. Then

$$k=\frac{\frac{1}{\mu}e^{-\beta(\varepsilon_0-\varepsilon_1)}+pe^{-\beta(\varepsilon_c-\varepsilon_1)}\left[\{(R_{0c}+R_{1c})+\frac{R_{1c}^2}{\mu RR_{10}}\}+\alpha\{1+\frac{e^{\beta(\varepsilon_c-\varepsilon_0)}+\frac{R_{1c}}{R_{10}}}{\mu R}\}\right]}{(1+e^{-\beta(\varepsilon_0-\varepsilon_1)})+\frac{1}{\mu}\frac{R_{1c}}{RR_{10}}}\ , \tag{11}$$

$$R=R_{0c}+R_{1c}+e^{-\beta(\varepsilon_c-\varepsilon_0)}\frac{R_{0c}R_{1c}}{R_{10}}\ .$$

If the Boltzmann distribution is assumed for $n_i(+\infty)$, which is the case of Gomer and Swanson,

$$k_{GS}=\frac{\frac{1}{\mu}e^{-\beta(\varepsilon_0-\varepsilon_1)}+pe^{-\beta(\varepsilon_c-\varepsilon_1)}\{(R_{0c}+R_{1c})+\alpha\}}{1+e^{-\beta(\varepsilon_0-\varepsilon_1)}}\ . \tag{12}$$

If μ approaches to $+\infty$, k approaches to k_{GS} as mentioned.

$$k=k_{GS}=\frac{pe^{-\beta(\varepsilon_c-\varepsilon_1)}\{(R_{0c}+R_{1c})\ \alpha\}}{1+e^{-\beta(\varepsilon_0-\varepsilon_1)}}\ . \tag{13}$$

If μ approaches to zero,

$$k=\frac{e^{-\beta(\varepsilon_0-\varepsilon_1)}+pe^{-\beta(\varepsilon_c-\varepsilon_1)}\{\frac{R_{1c}}{RR_{10}}+\frac{e^{\beta(\varepsilon_c-\varepsilon_0)}+\frac{R_{1c}}{R_{10}}}{R}\}}{\frac{R_{1c}}{RR_{10}}}\ , \tag{14}$$

which is independent of μ. On the other hand, k_{GS} becomes

$$k_{GS}=\frac{\frac{1}{\mu}e^{-\beta(\varepsilon_0-\varepsilon_1)}}{1+e^{-\beta(\varepsilon_0-\varepsilon_1)}}\ , \tag{15}$$

which depends on μ and diverges. From this example it is obvious that when μ is small, the redistribution effect is important and k_{GS} gives too large a rate of the ion desorption.

3. Microscopic Calculation of μ

In this section we perform a quantum formulation of the non-adiabatic process to calculate the mean first passage time μ. The basic Schrödinger equation ($\hbar=1$) for the wave function $\psi(r,R,t)$ is

$$i\frac{\partial}{\partial t}\psi(r,R,t)=\left[H_{el}(r,R)+T_R+V_R\right]\psi(r,R,t) \quad , \tag{16}$$

where r represents the coordinates of electrons and R represents the internal coordinates of nuclei, $H_{el}(r,R)$ stands for the electronic part of the total Hamiltonian, T_R and V_R are the nuclear kinetic and potential energy respectively. $\psi(r,R,t)$ is then expanded by the electronic wave functions $\phi_\alpha(r,R)$, and the nuclear wave functions $\chi_{\alpha i}(R)$,

$$\psi(r,R,t)=\sum_{\alpha i} C_{\alpha i}(t)e^{-i\varepsilon_{\alpha i}t}\chi_{\alpha i}(R)\phi_\alpha(r,R) \quad , \tag{17}$$

where α and i represent the electronic and nuclear state, and

$$[T_R+V_R+U_{\alpha\alpha}(R)]\chi_{\alpha i}(R)=\varepsilon_{\alpha i}\chi_{\alpha i}(R) \quad , \tag{18}$$

$$U_{\alpha\beta}(R)=\langle\phi_\alpha(r,R)|H_{el}(r,R)|\phi_\beta(r,R)\rangle \quad . \tag{19}$$

We choose $\{\phi_\alpha(r,R)\}$ as the orthonormal diabatic base. Then the time developing of the coefficient $C_{\alpha i}(t)$ is determined by

$$i\frac{\partial}{\partial t}C_{\alpha i}(t)=\sum_{\alpha\neq\beta}\sum_j C_{\beta j}(t)e^{i(\varepsilon_{\alpha i}-\varepsilon_{\beta j})t}\langle\chi_{\alpha i}(R)|U_{\alpha\beta}(R)|\chi_{\beta j}(R)\rangle \quad . \tag{20}$$

The diatomic molecule model of section 2 will be studied below. As we neglect the interaction between the vibrational states of the electronic state 1, only one bound state, i=0, is considered here. The continuum states of the electronic state 2 which is the ionic branch, are represented by k. The initial condition of this system is chosen as $C_0(0)=1$ and $C_k(0)=0$. Then the equations of $\{C_i\}$ are solved as follows:

$$C_0(t)=\frac{-e^{i\varepsilon_0 t}}{2\pi i}\int_{-\infty}^{\infty}d\omega G(\omega)e^{-i\omega t} \quad , \tag{21}$$

$$G(\omega)=\frac{1}{\left(\omega-\varepsilon_0-\sum_k\frac{|V_{0k}|^2}{\omega-\varepsilon_k+i0}\right)} \quad , \tag{22}$$

$$V_{0k}=\langle\chi_{10}|U_{12}|\chi_{2k}\rangle \quad . \tag{23}$$

$|C_0(t)|^2$ is the survival probability of the state 0 at time t, and related with the decaying rate of P(t) in section 2 by

$$P(t)=-\frac{d}{dt}|C_0(t)|^2 \quad . \tag{24}$$

Now the mean first passage time μ can be calculated by

$$\mu=\int_0^\infty dt|C_0(t)|^2=\frac{1}{2\pi}\int_{-\infty}^{+\infty}d\omega|G(\omega)|^2 \quad . \tag{25}$$

In the next step we must calculate the matrix element V_{0k}. As the potential surface of the state 2, the ionic branch, is almost a straight line, the nuclear wave function χ_{2k} can be described by an Airy function $\Phi(x)$,

$$\Phi(x)=\frac{1}{\sqrt{\pi}}\int_0^{+\infty} du \cos(xu+\frac{u^3}{3}) \ . \qquad (26)$$

Around the classical turning point of the potential curve of the state 1, the wave function χ_{10} may be approximated by also an Airy function. If we replace $U_{12}(R)$ by a constant U_0 around the crossing region, $|V_{0k}|^2$ becomes the square of an Airy function,

$$|V_{0k}|^2=U^2\Phi^2[-b(\frac{\varepsilon_0}{F_1}+\frac{\varepsilon_k}{F_2})] \ ,$$

$$U=\frac{|U_0|}{\sqrt{\pi}}(\frac{2mF_1F_2}{F_1+F_2})^{1/3} r_a d_0 \ , \quad b=(\frac{2mF_1F_2}{F_1+F_2})^{1/3} \ , \qquad (27)$$

where F_1 and F_2 are the slopes of the potential curves of the states 1 and 2 at each of the classical turning points, m is a reduced nuclear mass, r_a is a potential range of the electronic state 1, d_0 is a dimensionless scale parameter accompanying the approximation of the nuclear functions by Airy functions, and the origin of energy is chosen at the crossing energy. It is not easy to calculate $G(\omega)$ from this $|V_{0k}|^2$ form. As we are interested in the process around the crossing energy, $\Phi^2(x)$ is approximated by Lorentzian around x=0,

$$\Phi^2(x)\simeq\frac{1}{(x+1)^2+1} \ , \quad (x\sim 0) \ , \qquad (28)$$

then

$$|V_{0k}|^2\simeq\frac{a|V|^2}{(\varepsilon_k-\tilde{\varepsilon}_0)^2+a^2} \ , \qquad (29)$$

where

$$a=\frac{F_2}{b} \ , \quad |V|^2=aU^2 \ , \quad \tilde{\varepsilon}_0=\frac{F_2}{b}-\frac{F_2}{F_1}\varepsilon_0 \ .$$

Using above approximations, $1/\mu$ is calculated as

$$\frac{1}{\mu}=\frac{2(\frac{F_1F_2}{F_1+F_2})^2|U_0|^2d^2}{(\varepsilon_0-\varepsilon_T)^2+\varepsilon_T^2+\frac{F_1^3F_2^2}{(F_1+F_2)^3}\frac{|U_0|^2}{\varepsilon_T}d^2} \ , \quad d^2=\rho r_a^2 d_0^2 \ , \qquad (30)$$

$$\varepsilon_T=[\frac{1}{2m}(\frac{F_1F_2}{F_1+F_2})^2]^{1/3} \ , \qquad (31)$$

where ε_T is the energy scale of the tunneling region of the considered system, and ρ is the density of state k. The $|U_0|^2$ term of the denominator, which is missed in the simple perturbative calculation, plays an important role. When $|U_0|^2$ becomes large, the above expression gives finite $1/\mu$, on the other hand a perturbative calculation gives very large $1/\mu$ and too large transition probability.

4. Discussions

The rate of detection of ions k is obtained by (4),(7) and (25). Here we discuss only the qualitative features of k because we lack the real information about the energy levels of the neutral state (state 1). Roughly speaking, the rate k is the sums of $n(\varepsilon_i)$ times $1/\mu(\varepsilon_i)$. Then the activation energy Q is determined by the threshold energy, where $1/\mu$ changes most rapidly. If the change of $1/\mu$ is slow compared with the vibrational level spacing of the adsorbate, k consists of some exponential terms which cause the deviation from the simple Arrhenius form.

The rough sketches of the behaviors of $1/\mu$ to be expected in the cases of the diabatic and adiabatic limits are shown in Fig.3. In the case of the diabatic limit, the threshold energy almost coincides with the crossing energy. At the energy near the crossover, the quantum effect is enhanced and the change of $1/\mu$ with energy is smeared out. The lower energy region below the threshold represents the tunneling effect. Such behavior of $1/\mu$ results in the deviation of k from the simple Arrhenius relation, though the effective height of the activation barrier is not changed. In the case of the adiabatic limit, on the other hand, the threshold energy shifts to the peak energy of the adiabatic potential surface. The tail of $1/\mu$ into the lower energy region must be very small, since the tunneling effect is not significant. The behavior of $1/\mu$ in the general intermediate cases, on which we have concentrated our attention, have not been studied so far, although such might be crucially important for many non-adiabatic processes.

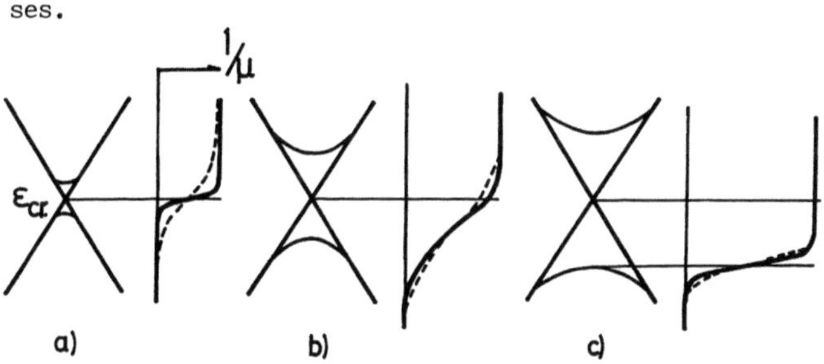

Fig.3 Schematic behavior of $1/\mu$ in a) diabatic, b) intermediate, c) adiabatic cases for small (solid line) and large tunneling (dotted line) rate

Equation (30) describes the diabatic limit and predicts some trends of the intermediate case. With the lowest perturbational approach such as done by Gomer and Swanson, the value of $1/\mu$ simply increases in proportion to the squared coupling constant $|U_0|^2$. By the result (30), however, $1/\mu$ saturates to a certain finite value for the large value of $|U_0|^2$. With the increase of the coupling constant $|U_0|^2$, the center of the changing region of $1/\mu$ shifts downwards and the width of the changing region increases as shown in Fig.3b. Based on such behavior of $1/\mu$, one might consider that the temperature-dependence of k cannot be described by the conventional Arrhenius relation. The shift of the center of the changing region of $1/\mu$ might bridge the gap between the adiabatic and the diabatic limit. The effect caused by such property of $1/\mu$ cannot be included in old theories. In the adiabatic limit, (30) cannot reproduce the expected rather sharp edge of $1/\mu$, because of the approximation (28), (29) for the Airy function. It should be remarked that the expression (30) of $1/\mu$ is not very good for the further region above the crossing energy.

The deviation of n(ε) from the Boltzmann factor becomes important when $\mu(\varepsilon)$ is small as discussed in section 2. It also causes the deviation from the simple activation form of k. Based on the relations (4),(7),(30), the field and temperature-dependence of ion detection rate k can be calculated for appropriate model systems. Such calculation will be a subject of further work.

In conclusion, it is important for the estimation of the ion detection rate of the field desorption to take into account the correct behavior of μ and the deviation from the Boltzmann population. To reproduce the correct magnitude and behavior of $1/\mu$, it is important to treat the interband tunneling by a non-perturbative method as proposed here.

References

1 E.W.Muller, T.T.Tsong:"Field Ion Microscopy, Field Ionization and Field Evaporation", in "Progress in Surface Science, Vol.4, Part 1", (Pergamon, Oxford, 1973)
2 O.Nishikawa, T.Yoshimura, M.Shibata:(unpublished)
3 O.Nishikawa, T.Yoshimura, M.Shibata: Surf. Sci. 124, 440 (1983)
4 Z.W.Gortel,H.J.Kreuzer, R.Teshima: Phys. Rev. B22, 5655 (1980)
5 R.Gomer, L.W.Swanson: J. Chem. Phys. 38, 1613 (1963)

Theory of Charge Exchange Collisions at Solid Surfaces

Ryoichi Kawai

Department of Physics, Waseda University, 3-4-1 Okubo
Shinjuku, Tokyo 160, Japan

1. Introduction

Theoretical studies of charge exchange scattering of an atom/ion at solid surface have been made for a long time. In early investigations, of which Hagstrum's works [1] are the most typical, ion-surface encounter is treated as quasistatic process and a neutralization rate is calculated on the basis of the so-called fixed ion approximation (FIA). However, there arise some problems. The FIA is not appropriate to the calculation of electron transfer probability involving levels above the Fermi level, for example, ionization of secondary ions [2], or excited states formation of hydrogen leaving from a metal surface [3]. Because these are essentially non-adiabatic phenomena, viz. electronic system experiences explicitly time-dependent pertubation as a result of the classical motion of the projectile. More precisely, quantum phase evolution plays an important role, which has been addressed by TRUBNIKOV and YAVLINSKII [4]. Recently, BLANDIN et al. [5], BLOSS and HONE [6], BRAKO and NEWNS [7], and MOYER and ORVEK [8] calculated the resonant neutralization probabilities of the low energy ions as a non-adiabatic process. Their investigations, however, are restricted to the very low energy case. It is the purpose of the present paper to calculate the resonant neutralization/ionization probabilities of the ion/atom with the intermediate energy (more than a few keV per nucleon) and to discuss some dynamical effects, e.g. electron translation factor [ETF], and energy uncertainty in consequence of the time-dependent pertubations which can not be found in the Hagstrum's formulae.

The formalism is developed in section 2, where a general formula to be compared with Hagstrum's is proposed. In section 3, dynamical effects are discussed in detail using a very simple model.

2. Formalism

The model used here is the same as that of BRAKO and NEWNS [7]. The moving atom is assumed to have a valence level Ψ_a of energy ε_a which interacts with the states Ψ_k of energy ε_k of the conduction band or the valence band. The nuclear motion is treated classically and separated from any electronic processes, so a hamiltonian of the electron system becomes time-dependent parametrically through the trajectory of the atom, and it may be written as

$$H=\varepsilon_a c_a^\dagger c_a+\sum_k \varepsilon_k c_k^\dagger c_k+\sum_k [V_k^*(t) c_a^\dagger c_k+V_k(t) c_k^\dagger c_a] \tag{2-1}$$

where $c_a^\dagger$, $c_k^\dagger$ are electron creation operators for states Ψ_a and Ψ_k. The energy ε_a is generally dependent on the position of the moving atom, but I neglect it for simplicity. If the atom moves straight with a velocity v, it may be written as following

$$\varepsilon_a=\varepsilon'_a+\tfrac{1}{2}mv^2, \qquad (2\text{-}2)$$

where ε'_a is the energy of the state Ψ_a in the inertial reference frames moving with a speed v along the path of the atom.

Using this hamiltonian, the Heisenberg equations of motion become

$$i\hbar\dot{d}_a=\sum_k V_k^*(t)\exp[-i(\varepsilon_k-\varepsilon_a)(t-t_o)/\hbar]d_k\ , \qquad (2\text{-}3)$$

$$i\hbar\dot{d}_k=V_k(t)\exp[i(\varepsilon_k-\varepsilon_a)(t-t_o)/\hbar]d_a\ , \qquad (2\text{-}4)$$

where we have defined

$$d_\nu(t)=\exp[i\varepsilon_\nu(t-t_o)/\hbar]c_\nu(t), \qquad (\nu=a,k) \qquad (2\text{-}5)$$

and time t_o refers to the system before the atom came into interaction with solid. Equation (2-4) is solved to give

$$d_k(t)=-(i/\hbar)\int_{t_o}^{t} V_k(t')\exp[i(\varepsilon_k-\varepsilon_a)(t'-t_o)/\hbar]d_a(t')dt' + d_k(t_o). \qquad (2\text{-}6)$$

And then, substituting (2-6) into (2-3), we obtain

$$i\hbar\dot{d}_a(t)=-(i/\hbar)\int_{t_o}^{t}\sum_k V_k^*(t)V_k(t')\exp[-i(\varepsilon_k-\varepsilon_a)(t-t')/\hbar]d_a(t')dt' +\sum_k V_k(t)\exp[-i(\varepsilon_k-\varepsilon_a)(t-t_o)/\hbar]d_k(t_o). \qquad (2\text{-}7)$$

Time variation of the system, on which energy uncertainty is based, is mainly caused by the motion of the swift atom. Therefore we may take out $d_a(t')$ from the integral in (2-7) putting t'=t, which I call the local time approximation (LTA). Accordingly, (2-7) becomes

$$i\hbar\dot{d}_a(t)=\Gamma(t)d_a(t)+\sum_k V_k(t)\exp[-i(\varepsilon_k-\varepsilon_a)(t-t_o)/\hbar]d_k(t_o), \qquad (2\text{-}8)$$

where

$$\Gamma(t)=-(i/\hbar)\int_{t_o}^{t}\sum_k V_k^*(t)V_k(t')\exp[-i(\varepsilon_k-\varepsilon_a)(t-t')/\hbar]dt'. \qquad (2\text{-}9)$$

$\Gamma(t)$ is the time-dependent self energy, whose imaginary part gives the effective transition rate discussed by MOYER and ORVEK [8] and whose real part means the time-dependent valence energy. This self-energy depends not only on events occuring at time t, but on the motion of the atom; in other words it approximately reflects a quantum phase evolution. In low energy case where oscillation of the neutral fraction can be seen[6], this approximation is not appropriate, because time evolution of $c_a(t)$ becomes important. It is worth while saying that OHTSUKI [9] has used the same kind approximation in calculating the local self-energy of the diffracted electron in crystals, and SHINDO [10] uses exactly the same approximation in derivation of the time-dependent Debye-Waller factor of the gas-surface inelastic collision. On the other hand, BRAKO and NEWNS [7] erased the history of the time evolution by means of the wide band limit, which is applicable in the very slow case.

Equation (2-8) may be solved to give

$$d_a(t)=-(i/\hbar)\int_{t_o}^{t}\sum_k V_k(t')\exp[-i(\varepsilon_k-\varepsilon_a)(t'-t_o)/\hbar]$$

$$\times\ \exp[-(i/\hbar)\int_{t'}^{t}\Gamma(\tau)d\tau]dt'd_k(t_o)$$

$$+\ \exp[-(i/\hbar)\int_{t_o}^{t}\Gamma(\tau)d\tau]d_a(t_o). \qquad (2\text{-}10)$$

We obtain the neutralization probability, which is given by the occupation numbers of the state Ψ_a , as following

$$n_a(t)=(2/\hbar^2)\mathrm{Re}\{\int_{t_o}^{t}dt_2\int_{t_o}^{t_1}\sum_k n_k(t_o)V_k^*(t_2)V_k(t_1)$$

$$\times\ \exp[-i(\varepsilon_k-\varepsilon_a)(t_1-t_2)/\hbar]\exp[-(i/\hbar)\int_{t_1}^{t_2}\Gamma(\tau)d\tau]$$

$$\times\ \exp[(2/\hbar)\int_{t_2}^{t}\mathrm{Im}\Gamma(\tau)d\tau]dt_1\}$$

$$+\ n_a(t_o)\exp[(2/\hbar)\int_{t_o}^{t}\mathrm{Im}\Gamma(\tau)d\tau]. \qquad (2\text{-}11)$$

Following the LTA, the second exponential function of the integrand should be taken unity. Finally, we obtain a simple formula of the neutralization probability as folowing

$$n_a(t)=n_a(t_o)\exp[-\int_{t_o}^{t}P(\tau)d\tau]+\int_{t_o}^{t}Q(t')\exp[-\int_{t'}^{t}P(\tau)d\tau]dt', \qquad (2\text{-}12)$$

where

$$P(t)=-(2/\hbar)\mathrm{Im}\Gamma(t)\quad ,\quad Q(t)=-(2/\hbar)\mathrm{Im}\Gamma'(t)\ , \qquad (2\text{-}13)$$

and $\Gamma'(t)$ is defined by

$$\Gamma'(t)=-(i/\hbar)\int_{t_o}^{t}\sum_k n_k(t_o)V_k^*(t)V_k(t')$$

$$\times\ \exp[-i(\varepsilon_k-\varepsilon_o)(t-t')/\hbar]dt'. \qquad (2\text{-}14)$$

Noting that (2-12) satisfies the following rate equation,

$$\dot{n}_a(t)=Q(t)-P(t)n_a(t)\ , \qquad (2\text{-}15)$$

we may conclude that P(t) corresponds to the hopping rate from atom to solid and Q(t) to that of the reversal process. It is the reason why P(t) does not equal Q(t) that electrons can hop above the Fermi level ε_f getting some energy from the atomic motion, which can not be seen in the Hagstrum's formula. Therefore, even though ε_a is above ε_f, it is possible for the atom to pick up an electron. In the next section we more precisely discuss it using a simple model.

3. Model Calculation

First we assume that the time variation of $V_k(t)$ does not depend on the value of k, so $V_k(t)$ may be written as

$$V_k(t)=V_k u(t). \tag{3-1}$$

Then (2-13) becomes

$$P(t)=(1/\hbar)\int d\varepsilon\Delta(\varepsilon)\int_{t_o}^{t} u(t)u(t')\cos[(\varepsilon-\varepsilon_a)(t-t')/\hbar]dt', \tag{3-2}$$

$$Q(t)=(1/\hbar)\int d\varepsilon\Delta(\varepsilon)f(\varepsilon,T)\int_{t_o}^{t} u(t)u(t')\cos[(\varepsilon-\varepsilon_a)(t-t')/\hbar]dt', \tag{3-3}$$

where

$$\Delta(\varepsilon)=(\pi/\hbar)\sum_k|V_k|^2\delta(\varepsilon-\varepsilon_k), \tag{3-4}$$

is the usual transition probability or the life time broadening,

$$f(\varepsilon,T)=\{1+\exp[(\varepsilon-\mu)/kT]\}^{-1}, \tag{3-5}$$

is the Fermi-Dirac function.

Roughly speaking, the results are characterized by an interaction time (a width of u(t) in time) and insensitive to the detailed time-variation of u(t). Because the time integrals in (3-2) and (3-3) give us a width of the energy uncertainty and nothing more, we choose

$$u(t)=\exp(-\gamma|t|), \qquad (\gamma=v/\alpha), \tag{3-6}$$

where α is a interaction length. Equation (3-6) is not only reasonable physically but convenient for integration. Then (3-2) and (3-3) become

$$P(t)=\int d\varepsilon\Delta(\varepsilon)g(\varepsilon)\{\exp(-2\gamma|t|) + 2[\cos((\varepsilon-\varepsilon_a)t/h)-\exp(-\gamma t)]\exp(-\gamma t)\theta(t)\}, \tag{3-7}$$

$$Q(t)=\int d\varepsilon\Delta(\varepsilon)g(\varepsilon)f(\varepsilon,T)\{\exp(-2\gamma|t|) + 2[\cos((\varepsilon-\varepsilon_a)t/h)-\exp(-\gamma t)]\exp(-\gamma t)\theta(t)\}. \tag{3-8}$$

where

$$g(\varepsilon)=\pi^{-1}\hbar\gamma/[(\hbar\gamma)^2+(\varepsilon-\varepsilon_a)^2], \tag{3-9}$$

gives a width of the energy uncertainty. The second terms of these equations may be neglected in our semiclassical case, because their oscillatory behavior reflects a quantum mechanical fluctuation and nothing more. However it must be said that this simplification will not change the qualitative feature of the results, but the oscillation is numerically not so small in the low energy case [11]. Consequently we obtain very simple expressions of the transition rates to be compared with the results of the FIA as following

$$P(t)=A|u(t)|^2, \quad A=\int \Delta(\varepsilon)g(\varepsilon)d\varepsilon, \tag{3-10}$$

$$Q(t)=A'|u(t)|^2, \quad A'=\int \Delta(\varepsilon)g(\varepsilon)f(\varepsilon,T)d\varepsilon. \tag{3-11}$$

It is not so unreasonable to choose a suitable $g(\varepsilon)$ for the other form of $u(t)$. The FIA offers to adopt the delta function as $g(\varepsilon)$, and BRAKO and NEWNS [7] have obtained another form

$$g(\varepsilon)=\mathrm{sech}[\pi(\varepsilon-\varepsilon_a)/\hbar\gamma]/\hbar\gamma. \tag{3-12}$$

In the low energy limit, both (3-9) and (3-12) become the delta functions. Finally (2-12) becomes

$$n_a(\infty)=(A'/A)[1-\exp(-A/\gamma)]. \tag{3-13}$$

Figure 1 gives the results of the numerical calculations for the resonant neutralization, in which it has been assumed that Δ is independent of ε, and (3-12) has been adopted instead of (3-9) because the range of the Lorentzian distribution is too wide. In the case where ε_a is between the Fermi level and the bottom of the conduction band (Fig.1-a), the LTA curves decrease with increasing velocity of the atom more rapidly than the FIA curves. In the other case where ε_a is above the Fermi level (Fig.1-b), the FIA does not allow an electron to hop above the Fermi level. On the other hand, the LTA neutralization probabilities are not so small and have a maximum in the dependence on the velocity. An application of the present result to the negative hydrogen formation is shown in ref.[11].

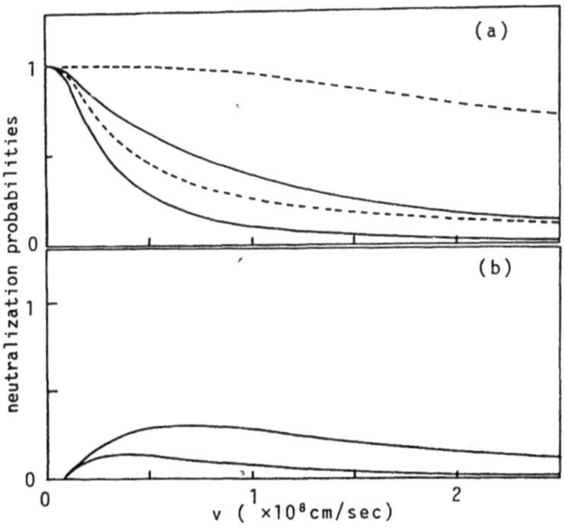

Fig.1 the neutralization probabilities: the solid lines show the LTA and the dashed lines the FIA for α =1A, T=0K, ε_f=5eV, ε_a =4eV (a), ε_a =6eV (b) from the band bottom, and Δ =10eV (upper curves), Δ =1eV (lower curves)

Acknowledgment

I should like to thank Prof. Y.H.Ohtsuki and Dr. S.Shindo for useful suggestions and discussions.

References

1. H.D. Hagstrum: Phys. Rev. **96**, 336 (1954)
   H.D. Hagstrum: in "inelastic ion-surface collisions", edited by N.H. Tolk, J.C. Tully, W. Heiland, and C.W. White (Academic, New York, 1977), p.1
2. Z. Šroubek, J. Zavadil, F. Kubec and K. Žďánský: Surface Sci. **77**, 603 (1978)
3. T. Gotoh, M. Kotani, Y. Kawaguchi, Y. Tazawa, Y. Shigeta and S. Ohtani: J. Phys. **D16**, 439 (1983)
4. B.A. Trubnikov and Yu.N. Yavlinskii: Soviet Phys.-JETP **25**, 1089 (1967)
5. A. Blandin, A. Nourtier and D.W. Hone: J. Physique **37**, 369 (1976)
6. W. Bloss and D. Hone: Surface Sci. **72**, 277 (1978)
7. R. Brako and D.M. Newns: Surface Sci. **108**, 253 (1981)
8. C.A. Moyer and K. Orvek: Surface Sci. **114**, 295 (1982)
9. Y.H. Ohtsuki: J. Phys. Soc. Japan **34**, 473 (1973)
10. S. Shindo: these proceedings
11. R. Kawai and Y.H. Ohtsuki: Nulear Instrum. & Methods **B2**, 414 (1984)

Energy Dissipation at Metal Surfaces: The Electron-Hole Pair Mechanism

K. Schönhammer

Institut für Theoretische Physik der Universität Göttingen
D-3400 Göttingen, Fed. Rep. of Germany

O. Gunnarsson

Max-Planck-Institut für Festkörperforschung
D-7000 Stuttgart, Fed. Rep. of Germany

1. Introduction

Recently there has been much interest in the energy-transfer between a surface and an atom or molecule approaching the surface. The energy-transfer processes e.g. determine the probability that the incoming particle loses its kinetic energy and is adsorbed at the surface. In the case of metal substrates the excitation of electron-hole pairs presents a mechanism in addition to the excitation of substrate phonons. We will consider here the electron-hole (eh) pair mechanism only. As we have reviewed the theoretical description of this mechanism recently [1], we will focus here on the problems one faces in the case when the inelasticity due to eh-pairs is strong.

A fully quantum mechanical despription of the electronic energy transfer mechanism can be carried out in a rather straightforward way only for the case of weak inelasticity [2-3]. This turns out to be a reasonable assumption for the interaction of rare gas atoms with metal surfaces [3]. To simplify the description for cases of strong inelasticity the adsorbate motion is often treated classically. A "thermal" atom approaching the metal surface then acts as a slowly varying localized perturbation of the electronic system.

A central quantity for the calculation of the sticking probability is the probability $P_{\varepsilon'}(\varepsilon)$ that an incoming atom (molecule) with energy ε' loses the energy ε during one round tip in the adsorption well. If the substrate temperature T_s is zero the sticking probability s is

$$s = \int_{\varepsilon'}^{\infty} P_{\varepsilon'}(\varepsilon)d\varepsilon$$

since the particle cannot leave the potential well if it has lost an energy $\varepsilon > \varepsilon'$. As the calculation of the full loss function $P_{\varepsilon'}(\varepsilon)$ is generally complicated, early efforts in the field [4,5] concentrated on the friction coefficient, which is directly related to the center of gravity of $P_{\varepsilon'}(\varepsilon)$. The friction coefficient is usually calculated assuming noninteracting electrons [4]. Using the Anderson model for chemisorption [6] some efforts have, however, been made to study intraatomic Coulomb effects on the friction coefficient [5,7]. The friction coefficient gives a good idea about the size of the nonadiabatic effects, but it does not, in general, give enough information about $P_{\varepsilon'}(\varepsilon)$ to calculate the sticking coefficient. In the following we will discuss the problems that occur in the calculation of the energy distribution function $P(\varepsilon)$ for a system of noninteracting electrons which is disturbed by a slowly varying perturbation of arbitrary strength. For "diagonal" perturbations [5] an asymptotically exact solution for the loss function can be obtained [8] by introducing coherent eh-pair excitation operators as first studied by Tomonaga [9] in a different context. For "nondiagonal" perturbations [5] a boson type approximation only produces the correct first moments μ_o, μ_1 and μ_2 with $\mu_n \equiv \int P(\varepsilon)\, \varepsilon^n\, d\varepsilon$. We have therefore studied a simple one-dimensional model for which the exact result for the loss function can be obtained numerically using techniques developed in the context of electron spectroscopy [10]. The relevance of these results for the calculation of the sticking coefficient and the concept of the friction coefficient is discussed.

2. The Energy Distribution Function

In the first part of this section we describe quite generally the effect of localized time-dependent perturbations in systems of noninteracting electrons. The Hamiltonian for the electrons in the time-dependent potential is given by

$$H_t = H_o + V_t \quad . \tag{2}$$

We assume that the perturbation V_t vanishes in the limit $t\to-\infty$ and approaches a time-independent operator (or zero) in the limit $t\to\infty$. We consider the case that the system is initially in its ground state $|E_o\rangle$. The generalization to finite temperatures T_s is rather straightforward [11,1]. We are interested in the probability that at $t\to\infty$ the system has made a transition to an excited state of the final Hamiltonian H_∞ due to the action of the perturbation. This is described by the energy distribution function $P(\varepsilon)$

$$P(\varepsilon) \equiv \langle E_o(\infty)\,|\,\delta(\varepsilon-(H_\infty - E_o^\infty))\,|E_o(\infty)\rangle \tag{3}$$

where $|E_o(\infty)\rangle = U(\infty,-\infty)\,|\,E_o\rangle$ is the solution of the time-dependent Schrödinger equation and E_o^∞ is the energy of the ground state of H_∞. Switching to the Heisenberg picture $P(\varepsilon)$ can be rewritten as

$$P(\varepsilon) = \langle E_o\,|\delta(\varepsilon-(H_\infty(\infty) - E_o^\infty))|\,E_o\rangle \tag{4}$$

where $H_t(t)$ is the Heisenberg operator

$$H_t(t) \equiv U^+(t,\infty)\,H_t\,U(t,\infty) \ . \tag{5}$$

Integrating the equation of motion for $H_t(t)$ one obtains

$$H_\infty(\infty) = H_o + \int_{-\infty}^{\infty} U^+(t,\infty)\,\dot{V}_t\,U(t,\infty)\,dt$$

$$\equiv H_o + W \tag{6}$$

where $\dot{V}_t$ is the time derivative of the Schrödinger operator V_t. If we replace $H_\infty(\infty)$ in (4) by $H_0 + W$ we are faced with a problem which is (apart from a constant shift) formally identical to the energy distribution function for the suddenly switched-on perturbation W. Once W is known, the calculation of $P(\varepsilon)$ therefore seems to be reduced to a problem solved in the context of x-ray photoemission from core levels [12]. The main difference, as will be discussed below, is the fact that for slow perturbations W is not a local perturbation, although it derives from the local perturbation V_t. The average energy transfer $\Delta\varepsilon \equiv \mu_1 = \int \varepsilon P(\varepsilon) d\varepsilon$ can be expressed as the average of the "perturbation" W

$$\Delta\varepsilon = \langle E_0|W|E_0\rangle + E_0 - E_0^\infty \tag{7}$$

where E_0 is the ground state energy of H_0. From the work on suddenly switched perturbations it is known that it is easier to study the Fouriertransform of $P(\varepsilon)$

$$P(\varepsilon) = \frac{1}{2\pi} \int_{-\infty}^{\infty} e^{i\varepsilon t} P(t) dt$$

with $$P(t) = \langle E_0 | e^{-i(H_0+W)t} | E_0 \rangle e^{iE_0^\infty t} \quad . \tag{8}$$

Various methods have been proposed to calculate P(t) [12,13]. In the time-dependent generalization of the exp (s) method [13] one obtains

$$P(t) = \exp \{-i\Delta\varepsilon \cdot t - i \int_0^t \sum_{\alpha,\mu} W_{\mu\alpha} S_{\alpha\mu}(t')dt'\} \tag{9}$$

where the $S_{\alpha\mu}(t)$ obey the nonlinear differential equations

$$i \dot{S}_{\alpha\mu}(t) = (\varepsilon_\alpha - \varepsilon_\mu) S_{\alpha\mu}(t) + W_{\alpha\mu} + \sum_\beta W_{\alpha\beta} S_{\beta\mu}(t) - \sum_\nu S_{\alpha\nu}(t) W_{\nu\mu} - \sum_{\beta\nu} S_{\alpha\nu}(t) W_{\nu\beta} S_{\beta\mu}(t) \tag{10}$$

and μ and α label states occupied in $|E_0\rangle$ while α and β label unoccupied states. The matrix elements W_{ij} (i,j occupied or unoccupied) are taken in the representation of the one-particle eigenstates $|\varepsilon_i\rangle$ of H_0

$$W_{ij} = \langle\varepsilon_i|W|\varepsilon_j\rangle = \int_{-\infty}^{\infty} \langle\varepsilon_i(t)|\dot{V}_t|\varepsilon_j(t)\rangle dt \tag{11}$$

where we have introduced the time dependent one-particle states $|\varepsilon_i(t)\rangle = U(t,\infty)|\varepsilon_i\rangle$. In our approach the calculation of $P(\varepsilon)$ has been split into two parts: The calculation of the W_{ij} and the subsequent calculation of P(t) using (9) and (10). As discussed below this splitting, which can be avoided using the Keldysh-formalism [14], is very convenient for the case of slow perturbations.

If W is assumed to be small one obtains from (10) to leading order in W

$$P(t) = \exp\{-\sum_{\alpha\mu} \frac{|W_{\alpha\mu}|^2}{(\varepsilon_\alpha-\varepsilon_\mu)^2}(1-e^{-i(\varepsilon_\alpha-\varepsilon_\mu)t})\} \tag{12}$$

and

$$\Delta\varepsilon = \sum_{\alpha,\mu} \frac{|W_{\alpha\mu}|^2}{(\varepsilon_\alpha-\varepsilon_\mu)} \quad . \tag{13}$$

This should be obviously a good approximation if V_t itself is small and varies over a finite time. Then (apart from an irrelevant phase factor) we can approximate the time-dependent states in (11) by the zeroth order expression $|\varepsilon_i(t)\rangle \approx \exp(-i\varepsilon t)|\varepsilon_i\rangle$. This leads to the result of Müller-Hartmann et. al. (MHRT) [15]

$$W_{ij}^{MHRT} = \int_{-\infty}^{\infty} \langle\varepsilon_i|\dot{V}_t|\varepsilon_j\rangle e^{i(\varepsilon_i-\varepsilon_j)t} dt \quad . \tag{14}$$

For a slowly varying perturbation V_t of <u>arbitrary</u> strength it nevertheless can happen that W is small. Then (13) should still be a good approximation but (14) cannot be used. For a slow perturbation and an infinite system it is useful to introduce the adiabatic one-particle states $|\varepsilon,\lambda\rangle^t$, where $H_t|\varepsilon,\lambda\rangle^t = \varepsilon|\varepsilon,\lambda\rangle^t$. Here we have labeled the states by the energy ε and some additional quantum numbers λ. Except for a rapidly varying phase factor, we can in this limit approximate the time-dependent one-particle states by $|\varepsilon,\lambda\rangle^t$

$$W_{\varepsilon\lambda,\varepsilon'\lambda'}^{BAI} = \int_{-\infty}^{\infty} {}^t\langle\varepsilon,\lambda|\dot{V}_t|\varepsilon',\lambda'\rangle^t \, e^{i(\varepsilon-\varepsilon')t} dt \tag{15}$$

which we have called the "boson approximation I" (BAI) [16]. For the case of degenerate one-particle states a unitary transformation within the subspace of the degenerate adiabatic states leads to a new set of adiabatic states. The question of the 'correct' choice of states is discussed later. Before we discuss the condition for W_{ij}^{BAI} to be small, we introduce one further simplification: As the matrix element in the integrand in (15) is slowly varying with time (the localized perturbation V_t is assumed to vary on a time scale T) the operator W has nonvanishing matrix elements W_{ij} only for $|\varepsilon_i - \varepsilon_j| \lesssim 1/T$. For a localized potential the matrix element ${}^t\langle\varepsilon,\lambda|\dot{V}_t|\varepsilon',\lambda'\rangle^t$ varies with the energies ε and ε' on the scale of typical electronic energies. It is then a good approximation to evaluate ${}^t\langle\varepsilon,\lambda|\dot{V}_t|\varepsilon',\lambda'\rangle^t$ with both energy variables at the Fermi energy ε_F ("boson approximation II" (BAII) [16])

$$W_{\varepsilon\lambda,\varepsilon'\lambda'}^{BAII} = \int_{-\infty}^{\infty} {}^t\langle\varepsilon_F,\lambda|\dot{V}_t|\varepsilon_F,\lambda'\rangle^t \, e^{i(\varepsilon-\varepsilon')t} dt \equiv \upsilon_{\lambda\lambda'}(\varepsilon-\varepsilon') \; . \tag{16}$$

As $\upsilon_{\lambda\lambda'}(\varepsilon-\varepsilon')$ is a function of the energy difference only, one of the energy integrations in (13) can be carried out. Using the normalization ${}^t\langle\varepsilon,\lambda|\varepsilon',\lambda'\rangle^t = \delta_{\lambda\lambda'}\,\delta(\varepsilon-\varepsilon')$ we obtain

$$P(t) = \exp\{-\sum_{\lambda,\lambda'}\int_0^\infty \frac{|\upsilon_{\lambda\lambda'}(\varepsilon)|^2}{\varepsilon}(1-e^{-i\varepsilon t})\,d\varepsilon\}. \tag{17}$$

The matrix elements ${}^t\langle\varepsilon,\lambda|\dot{V}_t|\varepsilon,\lambda'\rangle^t$ can be related to the scattering matrix $S^t(\varepsilon)$ corresponding to the static potential V_t. The S-matrix can be expressed in terms of the T-matrix in the standard way

$$\langle\varepsilon,\lambda|S_t|\varepsilon',\lambda'\rangle = \delta(\varepsilon-\varepsilon')\; S^t_{\lambda\lambda'}(\varepsilon)$$

$$= \delta(\varepsilon-\varepsilon')\,[\,\delta_{\lambda\lambda'} - 2\pi i\,T^t_{\lambda\lambda'}(\varepsilon+io)\,] \tag{18}$$

The 'correct' adiabatic states to be used in (16) are the (incoming wave) scattering states $|\varepsilon,\lambda+\rangle$

$$|\varepsilon,\lambda\rangle^t = |\varepsilon,\lambda+\rangle = \lim_{\eta\to 0}\frac{i\eta}{\varepsilon-H_t+i\eta}\,|\varepsilon,\lambda\rangle \tag{19}$$

The T-matrix is given by $T^t_{\lambda\lambda'}(\varepsilon+io) = \langle\varepsilon,\lambda|V_t|\varepsilon,\lambda'+\rangle$. Differentiation with respect to the time label leads to [1]

$${}^t\langle\varepsilon,\lambda|\dot{V}_t|\varepsilon,\lambda'\rangle^t = [\,(1-2\pi i T)^{-1}\,\frac{\partial T}{\partial t}\,]_{\lambda\lambda'}\quad . \tag{20}$$

Therefore weak inelasticity occurs if the scattering T-matrix is small and varies over a finite time. The important special case of "diagonal" perturbations results if the eigenstates of the unperturbed Hamiltonian can be chosen in such a way that V_t is diagonal in the additional quantum numbers λ. Then the T-matrix is also diagonal and (20) simplifies to

$${}^t\langle\varepsilon,\lambda|\dot{V}_t|\varepsilon,\lambda'\rangle^t = -\delta_{\lambda\lambda'}\;\dot{\delta}_{\varepsilon,\lambda}(t)/\pi \tag{21}$$

where we have introduced the phase-shifts $\delta_{\varepsilon,\lambda}(t)$ in the usual way by writing $S^t_{\lambda\lambda'}(\varepsilon) = \delta_{\lambda\lambda'}\,\exp(2i\delta_{\varepsilon,\lambda}(t))$. For this important special case (see next section) the result (17) for P(t) is asymptotically exact for slow perturbations even if the inelasticity is not weak. We have shown this [8] using coherent eh-pair excitation operators as first used by Tomonaga [9] in a different context. Unfortunately this statement is no longer true for "nondiagonal" perturbations V_t. As discussed previously [1,8] the result (17) for P(t) generally leads asymptotically to the correct first moment $\mu_1 = \Delta\varepsilon$ of $P(\varepsilon)$. The same holds true for the second moment. The exact result for the second cumulant $\varkappa_2 \equiv \mu_2 - \mu_1^2$ can be obtained from the short-time behaviour of $S_{\alpha\mu}(t)$ which follows from (10).

$$\varkappa_2 = \sum_{\alpha\mu} |W_{\alpha\mu}|^2 . \tag{22}$$

In the slow limit the simplification (16) is allowed and (22) agrees with the result for $\varkappa_2$ which follows from (17). The result for the third cumulant $\varkappa_3 \equiv \mu_3 - 3\mu_1\mu_2 + 2\mu_1^3$ follows from (10) as

$$\varkappa_3 = W_{\mu\alpha}(\varepsilon_\alpha - \varepsilon_\mu)W_{\mu\alpha} + W_{\mu\alpha}W_{\alpha\beta}W_{\beta\mu} - W_{\mu\alpha}W_{\alpha\nu}W_{\nu\mu} \tag{23}$$

where a summation over the indices is implied. In the slow limit (16) can be used and one obtains

$$\varkappa_3 = \int_0^\infty \omega^2\, \mathrm{Tr}(\upsilon(\omega))\, \upsilon(-\omega))\, d\omega$$

$$+ \int_0^\infty d\omega_1\, \omega_1 \int_{-\infty}^{\omega_1} \mathrm{Tr}\, \{\upsilon(\omega_2 - \omega_1)[\, \upsilon(-\omega_2), \upsilon(\omega_1)\,]\}\, d\omega_2 \tag{24}$$

where the trace is over the additional quantum numbers λ. The second term involving the commutator of two υ-matrices is not reproduced by the BAII in (17). Although this term vanishes for diagonal perturbations, (24) shows that (17) is generally not asymptotically exact for slow "nondiagonal" perturbations. A more serious shortcoming of (17) for nondiagonal perturbations is the fact that the elastic peak usually has zero weight even if the perturbation V_t is switched off in the limit $t \to \infty$. If there is an appreciable weight in the exact result for the elastic peak it is therefore necessary to go beyond the leading approximation for $S_{\alpha\mu}(t)$ in the solution of (10). This will be discussed in a forthcoming publication [17]. Examples for both types of perturbations are discussed in the next section.

3. Applications

The most studied case of a diagonal perturbation is a spherical potential in a homogenous electron gas [12]. In this case the perturbation is diagonal in the angular momentum quantum numbers l and m, when the angular momentum states $|\varepsilon, l, m\rangle$ are taken as the unperturbed states.

A simple description of the interaction of an adsorbate with a metal substrate is provided by the (spinless) Anderson model [6] in which only a single adsorbate level is taken into account. The Hamiltonian reads

$$H_t = \sum_k \varepsilon_k n_k + \varepsilon_a(t) n_a + \sum_k [V_{ak}(t)\, \phi_a^+ \phi_k + \mathrm{h.c.}] \tag{25}$$

where the substrate levels are described by ε_k, k labeling the wave vector $\vec{k}$ and the band index n, the adsorbate level by $\varepsilon_a(t)$ and $V_{ak}(t)$ gives the interaction between the two systems. The time dependence is determined by the classical trajectory $\vec{R}(t)$ of the adsorbate, e.g. $V_{ak}(t) = \tilde{V}_{ak}(\vec{R}(t))$. The description simplifies considerably, when the interaction is with the same local metal state all the time, and only the interaction strength varies with time, i.e. $V_{ak}(t) = V(t)\,A_k$. This may be a reasonable assumption for the case of normal incidence. Then the perturbation presented by the adsorbate couples only to a one-dimensional subspace of metal states

$$|\varepsilon> \equiv \sum_n \int d\vec{k}\, A_{\vec{k},n}\, \delta(\varepsilon-\varepsilon_{\vec{k},n})\, |\,\vec{k},n> \tag{26}$$

and we have the diagonal case with no additional quantum numbers. The interaction is then described by a single phase-shift $\delta_{\varepsilon_F}(t)$ [1,8]. Nørskov and Lundquist [18] have proposed that the creation of eh-pairs should be very important for adsorbates which have an affinity level close to the Fermi level. For the case when the adsorbate level crosses the Fermi level in a time Δt when approaching the surface and after a round trip time T crosses the Fermi level again from below, the time dependence of $\delta_{\varepsilon_F}(t)$ is qualitatively given by

$$\dot{\delta}_{\varepsilon_F}(t) = \frac{\Delta t/2}{t^2+(\Delta t/2)^2} - \frac{\Delta t/2}{(t+T)^2+(\Delta t/2)^2} \quad . \tag{27}$$

For this form of $\dot{\delta}_{\varepsilon_F}(t)$ the loss function $P(\varepsilon)$ can be calculated analytically [8]

$$P(\varepsilon) = \frac{(\Delta t)^2}{(\Delta t)^2+T^2}\,[\,\delta(\varepsilon) + T^2\varepsilon\, e^{-\varepsilon\,\Delta t}] \, . \tag{28}$$

It is easy to see that the width of the distribution is of the same order as the average energy-transfer, i.e. the probability for an energy-loss is not sharply peaked around the average energy-loss as tacitly assumed in the friction-coefficient approach to sticking. This is due to the fact that the phase-shift determining the exponent in (17) is bounded. This simple picture is a consequence of the factorization $V_{ak}(t) = V(t)\,A_k$. For glancing incidence of the adsorbate this approximation fails completely, as the adsorbate will interact with Wannier states centered on many different surface atoms. Then the adsorbate presents a "nondiagonal" perturbation and (17) does not provide the exact solution in the slow limit.

To understand the behaviour of the loss function $P(\varepsilon)$ for a nondiagonal perturbation we have studied a simple one-dimensional model with a nondiagonal perturbation, which allows an exact numerical calculation of $P(\varepsilon)$ [17]. We consider an infinite one-dimensional tight-binding s-band, with $|m>$ labeling the Wannier states and $|k>$ the Bloch states which diagonalize the unperturbed Hamiltonian. The lattice spacing a is taken as the unit length i.e. $a = 1$. The perturbation is chosen as separable

$$<k|V_t|k'> = V_0(t)\ <k|x(t)><x(t)|k'> \tag{29}$$

where $|x(t)>$ is a time-dependent local state and $V_0(t)$ is the interaction strength. With the choice

$$<k|x(t)> = \frac{1}{\sqrt{2\pi}}\, e^{-ikx(t)} \tag{30}$$

the state $|x(t)\rangle$ is just the Wannier state $|m\rangle$ if $x(t) = m$ and $|x(t)\rangle$ is a more extended object if $x(t)$ is not integer. The additional quantum number λ in this model is the <u>direction</u> in which the Bloch wave travels. If we choose as the unperturbed one particle states $|\varepsilon,\lambda\rangle \equiv c(\varepsilon)||k|,\lambda\rangle$ with $\lambda = \pm 1$ the perturbation (29) is nondiagonal in the λ indices. For nearest-neighbor hopping one obtains $|c(\varepsilon)|^2 = (B^2-\varepsilon^2)^{-1/2} \equiv \pi \rho_0(\varepsilon)$ if the unperturbed band extends from -B to B. The exact time dependent states $|k(t)\rangle = U(t,-\infty)|k\rangle$ have been obtained by the numerical integration of the Schrödinger equation for the $b_{kk'}(t) \equiv \langle k|k'(t)\rangle$

$$(i\frac{\partial}{\partial t} - \varepsilon_k)\, b_{kk'}(t) = \frac{V_o(t)}{2\pi}\, e^{-ikx(t)} \int_{-\pi}^{\pi} e^{ik''x(t)}\, b_{k''k'}(t)dk''. \quad (31)$$

As for slow perturbations mainly low-energy eh-pairs are created, we have chosen a nonuniform discretization of the wave vector with a smaller spacing near the Fermi points $\pm k_F$ [17]. The results shown below are for the half-filled band case $k_F = \pi/2$. Once W has been calculated $P(\varepsilon)$ can be calculated by a numerical integration of (10) or by another technique developed in connection with x-ray spectroscopy [12,10].

The adiabatic matrix elements determining the matrix $\psi_{\lambda\lambda'}(\varepsilon)$ (16) are given by

$$t_{\langle\varepsilon_F,\lambda|\dot{V}_t|\varepsilon_F,\lambda'\rangle}t = \frac{\rho_o(\varepsilon_F)}{2} \Big\{ -i\dot{x}(t)k_F V_o(t)\Big[\frac{\lambda}{1+i\pi V_o\rho_o(\varepsilon_F)} - \frac{\lambda'}{1-i\pi V_o\rho_o(\varepsilon_F)}\Big] + \frac{\dot{V}_o(t)}{1+(V_o\pi\rho_o(\varepsilon_F))^2} \Big\}\, e^{-ik_F(\lambda-\lambda')x(t)} \quad (32)$$

In Figs 1 and 2 we present some exact results for $P(\varepsilon)$ and compare them with the boson approximation (17). As the energy unit we take $B = 1$. Slow perturbations then involve times much larger than one. The perturbation strength $V_0(t)$ is switched on smoothly in a time $2t_0$ [17], stays constant for a time $t_{max}-4t_0$ and is then switched off "symmetrically" in the time $2t_0$. In Fig. 1 the velocity $\dot{x}(t)$ is switched on smoothly in the time $2t_0$, stays constant for a time $t_{max}/2 - 4t_0$ and is switched off symmetrically. From time $t_{max}/2$ till t_{max} the time reversed path $x(t)$ is taken. In Fig. 2 the velocity is not reversed and stays constant for a time $t_{max} - 4t_0$ until it is switched off. The strength of the perturbation is chosen rather large $(V_0)_{max} = \pi$, which produces a localized state well above the top of the band. The values of t_0, t_{max} and the maximum velocity $\dot{x}_{max}$ are given in the figure captions. In Fig. 1a the perturbation is not moved ($\dot{x}(t) \equiv 0$) and we have the case of a diagonal perturbation. The boson approximation (17) agrees with the exact solution within the accuracy of the drawing. This excellent agreement can also be seen from the results for the weights of the elastic peak in the exact fermion calculation ($p_o^{(F)}$) and the boson approximation ($p_0^{(B)}$). The oscillations are due to the fact that the perturbation only acts for the finite time t_{max}. In Figs. 1b and 1c the

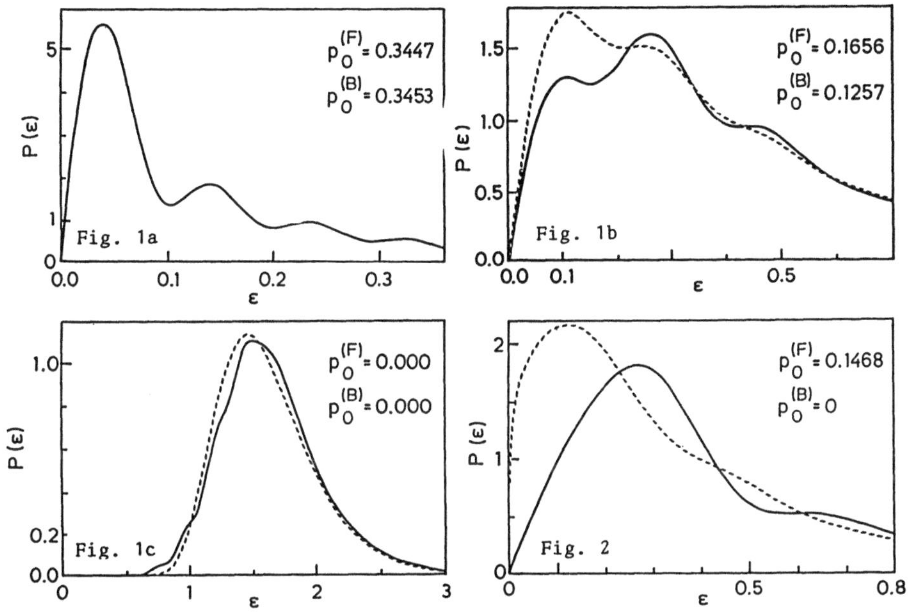

Fig. 1; Energy-loss function P(ε) for the 'nondiagonal' model described in the text: ——— exact result, ---- BAII . a) $t_0 = 10$, $t_{max} = 80$, $\dot{x}_{max} = 0$, b) $t_0 = 5$, $t_{max} = 40$, $\dot{x}_{max} = 0.075$; c) $t_0 = 5$, $t_{max} = 200$, $\dot{x}_{max} = 0.075$.

Fig. 2: P(ε) for the asymmetrical path described in the text: ——— exact result$^{\bullet}$, ---- BAII , $t_0 = 5$, $t_{max} = 30$, $\dot{x}_{max} = 0.075$

perturbation is moved with a velocity small compared to the Fermi velocity $v_F = 1$. In Fig. 1b the perturbation is only moved a distance of the order one lattice site (and back). There is still an appreciable weight in the elastic peak. Due to the symmetric form of the perturbation the BA also yields a nonvanishing value for p_0. As $p_0^{(F)}$ and $p_0^{(B)}$ differ appreciably the agreement of the results for P(ε) is not very good. In the case shown in Fig. 1c the perturbation is moved many lattice sites and the weight in the elastic peaks is less than 10^{-4}. Apart from the fine structure in the exact result which shows additional bumps separated by $2k_F \dot{x}_{max}$, the agreement between the two results is rather good. In this limit the loss function is "sharply" peaked around the average energy-loss and its gross features are properly described by $\Delta\varepsilon$ and $\varkappa_2$. This behaviour is to be expected from the central limit theorem if the assumption of the statistically independent excitation of many eh-pairs is justified.

The disagreement between the exact result and the BA becomes worse when the perturbation is not moved back to its initial position as shown in Fig. 2. Now $p_0^{(B)}$ vanishes and to compensate for this error the BA shows more weight at small energies than the exact result. For large values of V_0 it does not seem to be straightforward to find a simple improvement of (17) which yields a satisfactory answer for the situation in Fig. 2.

4. Conclusions

We believe that the results for the one-dimensional model of a nondiagonal perturbation presented in section 3 show the main qualitative features to be expected for realistic 3-dimensional models, e.g. in the description of inelastic scattering from surfaces using the Anderson Hamiltonian (24) without the simplification $V_{ak}(t) = V(t) \cdot A_k$. An exact numerical calculation of $P(\varepsilon)$ does not seem feasable in such a case because of the coupling to many continua. In the case of glancing incidence, when the adparticle interacts with very many localized metal states, the energy-loss will be strongly peaked around the average energy loss. This is the only regime in which the friction coefficient contains the essential information about $P(\varepsilon)$.

As discussed elsewhere [1] we expect the excitation of eh-pairs to be an important mechanism for light chemically-reactive adsorbates. We have shown in this paper that even when the classical trajectory assumption is used there is a regime for the case of strong inelasticity where so far no good approximation exists to calculate the energy-loss spectrum.

References

1. K. Schönhammer and O. Gunnarsson, Many-Body Phenomena at Surfaces, Ed. D. Langreth and H. Suhl (Academic Press, Orlando) p. 421 (1984)
2. G.P. Brivio and T.B Grimley, Surf. Sci. 89, 226 (1979); J. Phys. C10, 2351 (1977)
3. O. Gunnarsson and K. Schönhammer, Phys. Rev. B25, 2514 (1982)
4. E.G. d'Agliano, P. Kumar, W. Schaich and H. Suhl, Phys Rev. B11, 2122 (1975); A. Blandin, A. Nourtier and D.W. Hone, J. Phys. Paris 37, 369 (1976)
5. A. Nourtier, J. Physique 38, 479 (1977)
6. P.W. Anderson, Phys. Rev. 124, 41 (1961); D.M. Newns, Phys. Rev. 178, 1123 (1969)
7. A. Yoshimori and J.L. Motchane, J. Phys. Soc. Jap. 51, 6 (1982)
8. K. Schönhammer and O. Gunnarsson, Phys. Rev. B22, 1629 (1982); K. Schönhammer, Z.Phys. B45, 23 (1981)
9. S. Tomonaga, Prog. Theor. Phys. 5, 544 (1950)
10. K. Schönhammer and O. Gunnarsson, Surf. Sci. 89, 575 (1979)
11. K. Schönhammer and O. Gunnarsson, Phys. Rev. B24, 7084 (1981)
12. P. Nozières and C. de Dominicis, Phys. Rev. 178, 1097 (1969); K.D. Schotte and U. Schotte, Phys. Rev. 182, 479 ((1969); D. Langreth, Phys. Rev. B1, 471 (1970); M. Combescot and P. Nozières, J. Physique 32, 913 (1972); A. Kotani and Y. Toyozawa, J. Phys. Soc. Japan 37, 912 (1974)
13. K. Schönhammer and O. Gunnarsson, Phys. Rev. B18, 6606 (1978)
14. K. Makoshi, J. Phys. C16, 3617 (1983)
15. E. Müller-Hartmann, T.V. Ramakrishnan and G. Toulouse, Phys. Rev. B3, 1102 (1971)
16. O. Gunnarsson and K. Schönhammer, Phys. Rev. B25, 2503 (1982)
17. O. Gunnarsson and K. Schönhammer, to be published
18. J.K. Nørskov and B.I. Lundqvist, Surf. Sci. 89, 251 (1979)

Energy Dissipation Under Time-Dependent Local Perturbations

K. Makoshi

Department of Material Physics, Faculty of Engineering Science
Osaka University, Toyonaka, Osaka 560, Japan

1. Introduction

When an atom is moving near a metal surface, we often consider the motion of the atom in a given classical trajectory (trajectory approximation). The atom can be regarded as a source of a time-dependent perturbation to the electrons at the metal surface. Since the electron-hole pair excitation is possible with arbitrarily small energy in the metal, we expect the nonadiabatic process even when the atomic motion is slow [1-5].

The classic problem in this field is that of the threshold singularity in x-ray absorption spectroscopy [6] and in x-ray photoemission spectroscopy [7]. The absorption of x-ray leaves a core-hole behind leading to a sudden switching-on of the scattering potential acting on the electrons in the conduction band of the metal. The electron system relaxes slowly and asymptotically to the new ground state which is orthogonal to the initial ground state, resulting in the threshold singularity. The problem was solved essentially by NOZIÈRES and de DOMINICIS [6]. They relate the overlap between a state without the potential and that after a sudden change of the potential to a generalised Green function. They solve the resulting Dyson equation using the Muskhelishvili techniques [8]. The Dyson equation is converted to the singular integral equation to reflect the long-time behaviour of the Green function. The physics of the problem is explained by the so-called orthogonality catastrophe [9]. Further developements on the overlap integral have been made [10-12]. The most general result has been obtained by YAMADA and YOSIDA [12].

For the surface problem, we have to deal with the general time-dependence of the perturbation. MÜLLER-HARTMANN et al. [1] pointed out that the probability of retrieving the ground state after a gradual switching-on and then -off of the perturbation is not unity even if the perturbation is slow. SCHÖNHAMMER and GUNNARSSON [3] have shown, using the Tomonagon approach [13], that the energy dissipation probability is expressed by the *instantaneous* phase-shift at the Fermi surface. The identical result has been obtained by BRAKO and NEWNS [4] who work in the electronic coordinate directly. However, the latter approach has been questioned by SCHÖNHAMMER [14] for terminating the linked cluster expansion up to the second order. The present author [15] has recently provided a rigorous relation between the energy dissipation function and newly defined Green functions with a certain extention of the KELDYSH [16] Green function formalism which has been employed by BLANDIN et al. [17] for discussing the charge exchange in atom-surface collisions. He also applied the theory to the slow limit of perturbation in the noninteracting Wolff model employing the Muskhelishvili techniques. The scattering potential, however, is not necessarily separable, especially the time derivative of the *instantaneous* S-matrix may have the off-diagonal elements. Such effect has been partially incorporated into the theory of the friction coefficient by NOURTIER [18] for the noninteracting electron system and by YOSHIMORI and MOTCHANE [19] for the interacting electron system.

Here we look for the general form of the energy dissipation function applying the Muskhelishvili techniques for T=0 which has been considered so far to be the most reliable method for taking into account the long-time behaviour of the Green function, in other words the Fermi surface effect. Although the rigorous relation between the Green functions and the energy dissipation function holds even for interacting electron systems, we confine ourselves here to the noninteracting system. In the first part, we follow ref. [15] to get the exact expression of the energy dissipation function in terms of the Green functions and show how the Fermi surface reflects itself to the long-time behaviour of the Green function. In the second part, we apply the Muskhelishvili techniques in a matrix form for the slow limit of atomic motion. The analysis has not yet been completed, but we draw several results in the last part.

2. Exact Result

The Hamiltonian we are dealing with is

$$H(t) = H_0 + H'(t) , \tag{1}$$

$$H_0 = \sum_k \varepsilon_k c_k^+ c_k , \tag{2}$$

$$H'(t) = \sum_{k,k'} V_{kk'}(t) c_k^+ c_{k'} . \tag{3}$$

Here H_0 is the initial Hamiltonian which is diagonalised by an appropriate set of quantum numbers $\{k\}$, $H'(t)$ the time-dependent local perturbation with $V(-\infty)=0$. We assume that $V(t)$ is sufficiently smooth function of t ensuring the convergence of integrations evaluated in later expressions.

The quantity we want to calculate is

$$P(\varepsilon) = \langle\infty|\delta(E_0+\varepsilon-H(\infty))|\infty\rangle , \tag{4}$$

the probability for the electron system to be found in states with energy ε above the initial energy E_0 which is one of the eigenvalues of H_0 $(E_0|-\infty\rangle=H_0|-\infty\rangle)$, or its Fourier transform

$$\Pi(\tau) = \langle\infty|\exp[-i(H(\infty)-E_0)\tau]|\infty\rangle , \tag{5}$$

$$P(\varepsilon) = \int \frac{d\tau}{2\pi}\Pi(\tau)e^{i\varepsilon\tau} . \tag{6}$$

We first consider the problem in a finite interval of time $[t_-,t_+]$ and later take the limits $t_\pm \to \pm\infty$. Since $\exp[-iH(t_+)\tau]$ is a time evolution operator under the fixed Hamiltonian $H(t_+)$, $\Pi(\tau)$ is expressed by an overlap integral:

$$\Pi(\tau) = \langle t_+|t_+ +\tau\rangle \exp(-iE_0\tau) . \tag{7}$$

Equation (7) is a generalisation of the theory of NOZIÈRES and de DOMINICIS [6] for the general time-dependence of the perturbation.

With the time-evolution operator, we transfer to the interaction representation as

$$\Pi(\tau) = \langle t_-|\tilde{U}|t_-\rangle , \tag{8}$$

$$\tilde{U} = (T_-\exp[i\int_{t_-}^{t_+} H_{It'}(t')dt'])(T_+\exp[-i\int_{t_-}^{t_+} H_{It'+\tau}(t')dt']) . \tag{9}$$

$$H_{It}(t') = \sum_{k,k'} V_{kk'}(t) C_k^+(t') C_{k'}(t') . \tag{10}$$

Here $T_+(T_-)$ is the time-ordering operator as arranging field operators in descending order from the left (right). The difference of time τ between the matrix element V and the field operators comes out from the noncancellation of $\exp[\pm iH_0\tau]$.

We apply the procedure for getting the thermodynamic potential in terms of the temperature Green function [20], introducing the dimensionless coupling constant λ as $H'(t)\to\lambda H'(t)$. The result is

$$\Pi(\tau) = \exp[C(\tau)] \tag{11}$$

$$C(\tau) = -\int_0^1 d\lambda \int_{-\infty}^{\infty} dt \sum_{k,k'} [V_{kk'}(t+\tau) G_{k'k}^{++}(t,t^+;\lambda) - V_{kk'}(t) G_{k'k}^{--}(t^+,t;\lambda)] , \tag{12}$$

where $t^+ = t+i0$ and the Green functions $G^{\sigma\sigma'}$ $(\sigma,\sigma'=\pm)$ obey the Dyson equation in 2×2 matrix form:

$$\underset{\sim}{G} = \underset{\sim}{G}_0 + \underset{\sim}{G}_0 \underset{\sim}{\Sigma}_\lambda \underset{\sim}{G} , \tag{13}$$

$$\underset{\sim}{G}(t,t') = \begin{pmatrix} G^{++}(t,t') & G^{+-}(t,t') \\ G^{-+}(t,t') & G^{--}(t,t') \end{pmatrix} , \tag{14}$$

$$\underset{\sim}{\Sigma}_\lambda(t,t') = \lambda \begin{pmatrix} V(t+\tau) & 0 \\ 0 & -V(t) \end{pmatrix} \delta(t-t') . \tag{15}$$

Since $\tilde{U}$ consists of two time evolution operators, one for time increasing branch and another for decreasing branch, we have 2×2 matrix form of $\underset{\sim}{G}$. In (13), we assume that the multiplication rule of matrices is applied for the time branch and quantum numbers as well as the time. When we put $\tau=0$, $\underset{\sim}{G}$ reduces to the ordinary Keldysh Green function which does not contain the Fermi surface effect as pointed out by BLANDIN et al. [17]. The expressions (11)∿(13) are valid even for the interacting electron system, although the self-energy (15) becomes more complicated.

As far as we consider the noninteracting case, we can proceed further. We divide $\underset{\sim}{G}$ into two parts as done in the previous study [15] where we have dealt with the local Green functions only. Details of manipulations will be published elsewhere.

$$\underset{\sim}{G} = \underset{\sim}{G}_S + \underset{\sim}{G}_L , \tag{16}$$

$$\underset{\sim}{G}_S = G^r \begin{pmatrix} 1 & 0 \\ 1 & 0 \end{pmatrix} - G^a \begin{pmatrix} 0 & 0 \\ 1 & 1 \end{pmatrix} + L \begin{pmatrix} 0 & 0 \\ 1 & 0 \end{pmatrix} , \tag{17}$$

$$\underset{\sim}{G}_L = -i\underset{\sim}{G}_S \begin{pmatrix} 0 & N(1+iLN)^{-1} \\ N & 0 \end{pmatrix} \underset{\sim}{G}_S , \tag{18}$$

$$N_{kk'}(t,t') = n_k^0 \, \delta(t-t_-)\delta(t'-t_-)\delta_{kk'} . \tag{19}$$

Here N is the initial occupation (n_k^0) matrix and G^r, G^a and L satisfy the following integral equations:

$$G^r = G_0^r + G_0^r \lambda V_+ G^r , \tag{20}$$

$$G^a = G_0^a + G_0^a \lambda V_- G^a , \tag{21}$$

$$L = G^r \lambda (V_- - V_+) G^a , \tag{22}$$

with the convention for matrix multiplication and $V_+ \to V(t+\tau)$ and $V_- \to V(t)$. Here $G_0^{r/a}$ is retarded/advanced free Green function which does not contain the statistics:

$$\begin{pmatrix} G_{0k}^r(t-t') \\ G_{0k}^a(t-t') \end{pmatrix} = \begin{pmatrix} -\vartheta(t-t') \\ \vartheta(t-t') \end{pmatrix} i\exp[-i\varepsilon_k(t-t')] \ . \tag{23}$$

It is easy to see that the local Green functions given by summing up over k of G_0^r and G_0^a are significant only in a range of time smaller than, or comparable to, the electronic time scale i.e. the inverse band width 1/D [15]. The Fermi surface effect is included in the infinite series $(1+iLN)^{-1}$. The cut-off of the occupation number results in the long-time behaviour. As seen from (20) and (21), G^r (G^a) inherits the step function $\vartheta(t-t')$ ($\vartheta(t'-t)$) from G_0^r (G_0^a), thus giving vanishing contribution to $C(\tau)$.

The long-time part, $G_L^{\sigma\sigma}$, is written from (18) as

$$G_L^{++}(t,t') = iG^rN(1+iLN)^{-1}(G^a-L), \tag{24}$$

$$G_L^{--}(t,t') = i(G^r+L)(1+iLN)^{-1}G^a \ . \tag{25}$$

When $\tau=0$ ($L=0$), (24) reduces to the expression given by BLANDIN et al. [17] for t'>t. When we recognise that the integration over t after taking the equal time limits is expressed as the trace of the matrix and

$$\partial G^r/\partial\lambda = G^rV_+G^r \tag{26}$$

etc., we can perform the integration over λ as

$$C(\tau) = \mathrm{Tr}_{\{k,t\}}\ell n(1+iLN) \ . \tag{27}$$

Here the variables on which we have to take trace are indicated in the curly bracket. This expression is very simple and seems to be closely related to BRAKO and NEWNS [4] method by the use of the linked cluster expansion. A slightly different but essentially same expression has been obtained recently by NOURTIER [21]. It enables us a systematic approach like $\dot{V}(t)$ expansion. However, it contains a difficulty to deal with the infinite series, as pointed out by SCHÖNHAMMER [14] for the results of BRAKO and NEWNS [4], although the Muskhelishvili method with (24) and (25), which is discussed in the next section, copes with this difficulty.

3. Slow Perturbation and Muskhelishvili Techniques

We consider first the slowness approximation for $G^r(t,t_-)$ following BRAKO and NEWNS [4]. The existence of N after G^r allows us to evaluate G^r only in such arguments of time. We expand G^r in infinite series:

$$G^r = G_0^r + G_0^r\lambda V_+G_0^r + G_0^r\lambda V_+G_0^r\lambda V_+G_0^r + \cdots \ . \tag{28}$$

In this expression, we have terms such as

$$W \equiv \sum_k \int dt_2 V_{qk}(t_1+\tau)G_{0k}^r(t_1-t_2)V_{kq'}(t_2+\tau)G_{oq'}^r(t_2-t_-) \ . \tag{29}$$

Since we sum over all k, the range $|t_1-t_2| \lesssim 1/D$ is of major contribution for (29) if V is not singular around the Fermi surface. For the slow perturbation compared with the electronic time scale, we replace $V(t_2+\tau)$ by $V(t_1+\tau)$ and perform the integration on t_2:

$$W \simeq \sum_k V_{qk}(t_1+\tau)g^r_{0k}(\varepsilon_{q'})V_{kq'}(t_1+\tau)G^r_{0q'}(t_1-t_-) \ , \qquad (30)$$

where $g^r_{0k}(\varepsilon)$ is the Fourier component of $G^r_{0k}(t-t')$:

$$g^r_{0k}(\varepsilon) = (\varepsilon-\varepsilon_k+i0)^{-1} \ . \qquad (31)$$

We make this approximation successively and obtain G^r as

$$G^r(t,t_-;\lambda) = \int \frac{d\varepsilon}{2\pi i}\,[1+g^r_0(\varepsilon)T_\lambda(\varepsilon;t+\tau)][g^a_0(\varepsilon)-g^r_0(\varepsilon)]G^r_0(t-t_-) \ . \qquad (32)$$

Here $T_\lambda(\varepsilon;t+\tau)$ is the *instantaneous* T-matrix satisfying

$$T_\lambda(\varepsilon;t+\tau) = \lambda V(t+\tau)[1-\lambda g^r_0(\varepsilon)V(t+\tau)]^{-1} \ . \qquad (33)$$

The expression (32) is the extension of that obtained by the author [15] for the local Green function as the *spectral representation*. Similar approximations are applied to G^a and L and we take the long-time approximation [15]. We obtain

$$G^{++}_L = -[1+g^r_0T_\lambda(t+\tau)]F_\lambda(t,t')S^\dagger_\lambda(t')[1+T_\lambda(t'+\tau)g^r_0] \ , \qquad (34)$$

$$G^{--}_L = -[1+g^a_0T^\dagger_\lambda(t)]S_\lambda(t+\tau)F_\lambda(t,t')[1+T^\dagger_\lambda(t')g^a_0] \ , \qquad (35)$$

where all *instantaneous* T-matrices and $g^{r/a}_0$ are those for the Fermi energy (ε=0) as well as the *instantaneous* on-shell S-matrix:

$$S_{\lambda kk'}(t) = [\delta_{kk'}-2\pi i\delta(\varepsilon_k-\varepsilon_{k'})T_{\lambda kk'}(t)]_{\varepsilon_k=0} \ . \qquad (36)$$

The common function $F_\lambda(t,t')$ obeys the following equation:

$$F_\lambda(t,t') = \frac{1}{t-t'+i0} + \int \frac{dt_1}{2\pi i}\,\frac{[S^\dagger_\lambda(t_1)S_\lambda(t_1+\tau)-1]F_\lambda(t_1,t')}{t-t_1+i0} \ . \qquad (37)$$

This is the desired form for applying the Muskhelishvili techniques. The way of solving the singular integral equation in the matrix form has been recently given by YAMADA and YOSIDA [12]. We present here the result for the nonadiabatic contributions only:

$$\begin{aligned}C_{NA}(\tau)=-\int_{-\infty}^{\infty}\frac{dt}{4\pi i}\int_0^1 d\lambda\,\mathrm{Tr}\Big\{&\frac{\partial\Delta}{\partial\lambda}\Delta^\dagger \sum_{n=1}^{\infty}\int \frac{dt_1\cdots dt_n}{(2\pi i)^n}\,\frac{[\Delta^\dagger_1-1]\cdots[\Delta^\dagger_n-1]}{(t_1-t+i0)\cdots(t_n-t_{n-1}+i0)(t_n-t+i0)}\\ &+\Delta^\dagger\frac{\partial\Delta}{\partial\lambda}\sum_{n=1}^{\infty}\int \frac{dt_1\cdots dt_n}{(2\pi i)^n}\,\frac{[1-\Delta_1]\cdots[1-\Delta_n]}{(t_1-t-i0)\cdots(t_n-t_{n-1}-i0)(t_n-t-i0)}\\ &+S_\lambda(t)S^\dagger_\lambda(t)\,\frac{\partial S_\lambda(t+\tau)}{\partial\lambda}S^\dagger_\lambda(t+\tau)-S_\lambda(t+\tau)S^\dagger_\lambda(t+\tau)\frac{\partial S_\lambda(t)}{\partial\lambda}S^\dagger_\lambda(t)\Big\}, \qquad (38)\end{aligned}$$

with abbreviations: $\Delta\equiv S^\dagger_\lambda(t)S_\lambda(t+\tau)$ and $\Delta_m\equiv S^\dagger_\lambda(t_m)S_\lambda(t_m+\tau)$.

If all matrices in (38) are commutative, we can perform the integration over λ, introducing the Fourier transform of $\ell nS(t)$:

$$C_{NA}(\tau)=\pi^{-2}\int_0^\infty d\omega\omega\,\mathrm{Tr}\{D(\omega)D(-\omega)\}(e^{-i\omega t}-1), \qquad (39)$$

$$2iD(\omega)\equiv \int dt\ e^{i\omega t}\ell nS(t). \qquad (40)$$

An example is the case when the S-matrix is written by the single phase-shift as in the previous works [3,5,15]. Then $D(\omega)$ is the Fourier transform of the *instantaneous* phase-shift at the Fermi energy. Another case in which (39) is applicable is the x-ray problem where $S(t)\equiv\vartheta(t)S$ i.e. $S(t)$ is factorised into its time-dependent scalar function and t-independent S-matrix. We get the long-time asymptote of the overlap integral as

$$\Pi(\tau)\sim(iD\tau)^{-\alpha}, \tag{41}$$

$$\alpha=-(4\pi^2)^{-1}\mathrm{Tr}\{\ln^2 S\}, \tag{42}$$

which leads to the threshold singularity: $P(\varepsilon)\sim\varepsilon^{\alpha-1}$ except for the relaxation shift.

For the general time-dependence of $S(t)$, the integration over λ has not been performed yet. However, we can draw some information from (38). The mean energy loss is calculated from the last two terms in (38):

$$\Delta\varepsilon=i\dot{C}_{NA}(0)=-\int_{-\infty}^{\infty}\frac{dt}{4\pi}\,\mathrm{Tr}\{\dot{S}^{\dagger}(t)S(t)\}^2, \tag{43}$$

which reduces to the result of NOURTIER [18] describing the friction coefficient for the spherical scattering potential.

The Debye-Waller factor: $P_0=\Pi(\infty)$ is calculated from the first two terms in (38). For these terms, the commutators between $\Delta(t)$'s at different times seem to be negligible for slow perturbation limit, then

$$P_0 = \exp(-\int_0^{\infty}d\omega\omega\mathrm{Tr}\{D(\omega)D(-\omega)\}/\pi^2). \tag{44}$$

The justification of (44) must be done by evaluating correction terms with various commutators of $\Delta(t)$'s. This has not been done yet.

4. Concluding Remarks

We have shown that the energy dissipation function $P(\varepsilon)$ can be expressed in terms of the newly defined Green functions . Equation (12) is valid even for the interacting electron system while (24), (25) and (27) only for the non-interacting system. These equations can be the basis for investigating the energy dissipation problem in future such as the effect of finite band width and that of the electron-electron interaction.

In the last half part, we attempted to obtain results for the slow limit. However, the noncommutativity of S-matrix at different times prevents us to obtain a simple expression of $C_{NA}(\tau)$. The use of (27) may provide us some information. Although, we may eventually obtain, for the single phase-shift, the identical result to that obtained by BRAKO and NEWNS [4] by expanding up to the second order of L, we have to clarify its relation to the result obtained by applying the Muskhelishvili techniques, which cope with the long-time behaviour. This study is now under way.

An alternative way might be the numerical calculation in a certain model system [22,23] while it becomes more elaborate as the perturbation becomes slower. In this case, we should be careful about the effect of the finite band width.

References

1. E. Müller-Hartmann, T.V. Ramakrishnan and G. Toulouse: Solid State Commun. 9, 99 (1971) ; Phys. Rev. B3, 1102 (1971)
2. J.K. Nørskov and B.I. Lundqvist: Surf. Sci. 89, 251 (1970)
3. K. Schönhammer and O. Gunnarsson: Phys. Rev. B22,1629 (1980)
4. R. Brako and D.M. Newns: Solid State Commun. 33, 713 (1980); J. Phys. C14, 3065 (1981)
5. P. Minnhagen: J. Phys. C15, 2293 (1982)
6. P. Nozières and C.T. de Dominicis: Phys. Rev. 178, 1097 (1969)
7. S. Doniach and M. Šunjić: J. Phys. C3, 285 (1970); S. Doniach: Phys. Rev. B2, 3898 (1970)
8. N.I. Muskhelishvili: Singular Integral Equations , edited by J.R.M. Radok (P. Noordhoff N.V., Gronigen 1953)
9. P.W. Anderson: Phys. Rev. Lett. 18, 1049 (1967)
10. D.R. Hamann: Phys. Rev. Lett. 26, 1030 (1971)
11. K. Yamada and K. Yosida: Prog. Theor. Phys. 59, 1061 (1978); Prog. Theor. Phys. 60, 353 (1978)
12. K. Yamada and K. Yosida: Prog. Theor. Phys. 62, 363 (1979); Prog. Theor. Phys. 68, 1504 (1982)
13. S. Tomonaga: Prog. Theor. Phys. 5, 544 (1950)
14. K. Schönhammer: Z. Phys. B45, 23 (1981)
15. K. Makoshi: J. Phys. C16, 3617 (1983)
16. L.V. Keldysh: Sov. Phys. - JETP 20, 1018 (1965)
17. A. Blandin, A. Nourtier and D.W. Hone: J. Physique 37, 369 (1976)
18. A. Nourtier: J. Physique 38, 479 (1977)
19. A. Yoshimori and J-L. Motchane: J. Phys. Soc. Jpn 51, 1826 (1982)
20. A.A. Abrikosov, L.P. Gorkov and I.E. Dzyaloshinskii: Method of Quantum Field Theory in Statistical Physics (Prentice-Hall, Inc., Englewood Cliffs)
21. A. Nourtier: to be published
22. K. Schönhammer and O. Gunnarsson: in this volume
23. J.E. Inglesfield: Surf. Sci. 127, 555 (1983)

Effects of Intraatomic Coulomb Interaction in the Time-Dependent Newns-Anderson Model

A. Yoshimori, K. Makoshi, and H. Kawai

Department of Material Physics, Faculty of Engineering Science
Osaka University, Toyonaka, Osaka 560, Japan

Effects of Intraatomic Coulomb interaction are investigated in the time-dependent Newns-Anderson model with the use of the time-dependent Hartree-Fock approximation. The simplest time-dependence is chosen for parameters in the Hamiltonian and nonadiabatic effects are examined when the system crosses the boundary between the "nonmagnetic and magnetic" regions. Some asymmetry in spin space is inevitable for the system to cross the boundary in the present Hartree-Fock approximation. Then time-dependence corresponding to a slow limit is assumed for the parameters and energy dissipation is investigated.

1. Introduction

There have been numerous theoretical investigations on nonadiabatic electronic processes due to atoms or molecules moving near metal surfaces during the past ten years or so. Many of these have used the Newns-Anderson model, whose Hamiltonian has time-dependent parameters [1-8]. However, effects of the intraatomic Coulomb interaction in this model have never been studied, though the problem of this interaction is recognized to be important.

The purpose of this article is to examine the effects of the intraatomic Coulomb interaction in the time-dependent Newns-Anderson model with use of the Hartree-Fock approximation. The time-dependent expectation values of the electron occupation numbers, $n_\sigma(t)$, are calculated in an orbital on a moving atom or molecule. (In the following we refer to an atom, but it can be a molecule). Since the time-dependent Hartree-Fock approximation is used, $n_\sigma(t)$ are included in the starting Hamiltonian and must be self-consistent. They are key quantities in this approximation. Having them we can calcaulate physical quantities. They also describe the nonadiabtic charge-transfer between the atom and the metal surface. By using the obtained $n_\sigma(t)$, the energy dissipation due to the electronic excitation is calculated by the standard formula in the slow limit [6-8].

The time-dependence of parameters in the Hamiltonian of the model is considered to be given. The so-called trajectory approximation gives explicit time-dependence of those parameters when the motion of the atom is given. In this article we take examples of simple time-dependence without referring to details of the motion. The wide band limit [1] is assumed also for simplicity.

In the next section, the model is described. The behaviors of $n_\sigma(t)$ are discussed in the simplest cases of the strong intraatomic interaction with the initial condition of incoming bare ions [9] in the section 3. In the section 4, $n_\sigma(t)$ are calculated with the initial condition of incoming neutral atoms in the slow limit. Then the energy dissipation is calculated. Summary and discussion is given in the last section.

2. Model

The Hamiltonian of the model is given by

$$H = \Sigma\varepsilon_k \hat{n}_{k\sigma} + \Sigma\varepsilon_{a\sigma}(t)\hat{n}_{a\sigma} + \Sigma[V_k(t)c^\dagger_{k\sigma}c_{a\sigma} + \mathrm{h.c.}] + U(t)\hat{n}_{a\uparrow}\hat{n}_{a\downarrow}, \tag{2.1}$$

where $c_{k\sigma}$ and $c_{a\sigma}$ are annihilation operators in the conduction electron state of a semi-infinite metal with "wavevector" k and spin σ and in the orbital of the moving atom, respectively, the origin of the energies ε_k and $\varepsilon_{a\sigma}$ is chosen as the Fermi energy, $\hat{n}=c^\dagger c$, $V_k(t)$ the admixture matrix element between the conduction electron state and the orbital of the moving atom, and $U(t)$ the intraatomic Coulomb interaction. We assume that the energy level of the orbital of the moving atom is spin-dependent, that is, $\varepsilon_{a\sigma}(t)=\varepsilon_a(t)-\sigma\varepsilon_m$ (σ, here, means $\pm$ for $\uparrow$ and $\downarrow$, respectively). The reason for this will be made clear in the next section.

We use the Hartree-Fock approximation for the interaction term, and have

$$H = \Sigma\varepsilon_{k\sigma}\hat{n}_{k\sigma} + \Sigma\varepsilon_\sigma(t)\hat{n}_{a\sigma} + \Sigma[V_k(t)c^\dagger_{k\sigma}c_{k\sigma} + \mathrm{h.c.}], \tag{2.2}$$

with

$$\varepsilon_\sigma(t) = \varepsilon_{a\sigma}(t) + U(t)n_{-\sigma}(t), \tag{2.3}$$

where $n_\sigma(t)=\langle\hat{n}_{a\sigma}(t)\rangle$ and $\hat{n}(t)$ is the operator in the Heisenberg representation. These $n_\sigma(t)$ describe probabilities of charge states of the atom within the present approximation as follows:

$$I^+(t) = 1-n_\uparrow(t)-n_\downarrow(t)+n_\uparrow(t)n_\downarrow(t), \tag{2.4}$$

$$I^0_\sigma(t) = n_\sigma(t)-n_\uparrow(t)n_\downarrow(t), \tag{2.5}$$

$$I^-(t) = n_\uparrow(t)n_\downarrow(t), \tag{2.6}$$

where $I^+(t)$, $I^0_\sigma(t)$ and $I^-(t)$ are the fraction of positive ion, neutral atom with spin σ and negative ion, respectively, at time t.

The expression of $n_\sigma(t)$ has been obtained in the wide band limit by BLANDIN, NOUTIER and HONE [1]. We show the expression,

$$n_\sigma(t) = i\pi^{-1}\int_{-\infty}^{t} dt_1 \int_{-\infty}^{t} dt_2 u(t_1)u(t_2)\exp[i\int_{t_2}^{t_1} dt_3\varepsilon_\sigma(t_3) - \int_{t_1}^{t} dt_3 u(t_3)^2$$
$$- \int_{t_2}^{t} dt_3 u(t_3)^2]/(t_1-t_2+i0^+) + n_\sigma(-\infty)\exp[-2\int_{-\infty}^{t} dt_3 u(t_3)^2], \tag{2.7}$$

where $V_k(t)$ is assumed to be $V_k u(t)$, $\pi\Sigma_k|V_k|^2\delta(\varepsilon-\varepsilon_k)$ independent of ε and denoted by Δ, time is in units of $\hbar/\Delta$, and energy in units of Δ. Equations (2.7) are a set of nonlinear integral equations for $n_\uparrow(t)$ and $n_\downarrow(t)$. If $n_\uparrow(t)=n_\downarrow(t)$, we call the nonmagnetic solution, and if $n_\uparrow(t)\neq n_\downarrow(t)$, the magnetic solution, following those in the equilibrium case. In the following we solve (2.7) numerically.

3. Sudden Switch-on and -off Case

The case of sudden switch-on and -off is discussed in this section. This may be the case of the simplest time-dependence. $u(t)$ is assumed to be the step function $\vartheta(t)$, while ε_a and U are assumed to be constant. The initial condition is given by $n_\sigma(0)=0$. This corresponds to a simplification that the incident atom experiences fields with average parameters during its stay at the metal surface. Since the sudden switch-off is assumed and after it no admixture exists, $n_\sigma(t)$ keep the values at the switch-off. Therefore it is enough to follow $n_\sigma(t)$ after the sudden switch-on. Parts of the results in this case have been reported already [9]. We present here some more details to supplement as well as its brief summary.

In Fig. 1 (a) and (b) the results of numerical calculation are shown. The symmetric case of $U=16$ and $\varepsilon_a=-8$, which corresponds to a strong magnetic case in equilibrium ($U>\pi$ is the condition for the magnetic solution in equilibrium for the symmetric case) is shown in Fig. 1(a) for $\varepsilon_m=0.01$. We have to have some parameter asymmetric in spin space, which is ε_m here, because otherwise everything is symmetric in spin space in the present case and the system never goes to a magnetic solution. The solid lines represent $n_\sigma(t)$, the dashed line $\Sigma_\sigma n_\sigma(t)$, the dotted line $\Sigma_\sigma I_\sigma^0(t)$, and the chain line $I^-(t)$. In Fig. 1(b), $n_\sigma(t)$ for $\varepsilon_m=0.001$, 0.01 and 0.1 are shown in the case of the same values of ε_a and U.

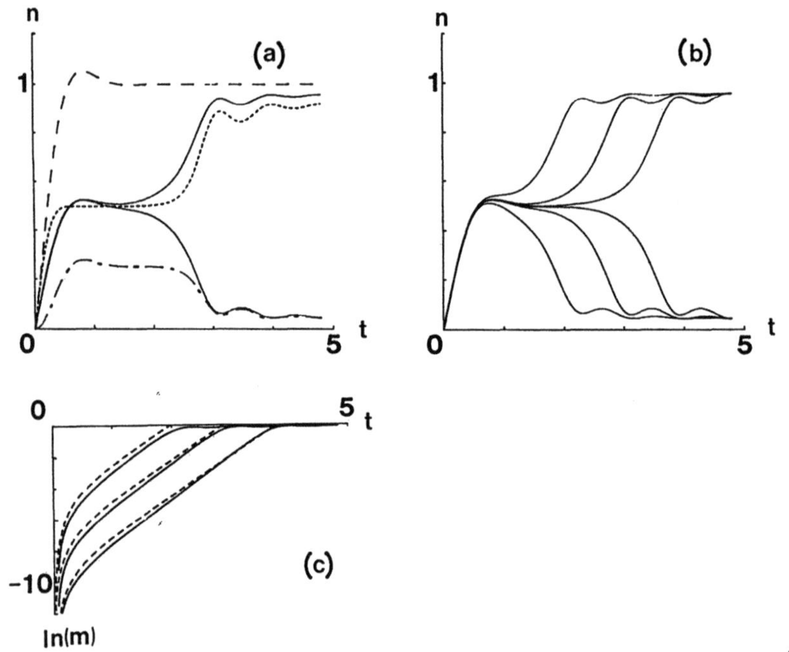

Fig. 1 (a) Curves of $n_\sigma(t)$ (solid line), $\Sigma_\sigma n_\sigma(t)$ (dashed line), $\Sigma_\sigma I_\sigma^0(t)$ (dotted line), $I^-(t)$ (chain line) in the case of $\varepsilon_a = -8$, $U=16$ and $\varepsilon_m=0.01$. (b) Curves of $n_\sigma(t)$ in the cases of $\varepsilon_m=0.001$, 0.01 and 0.1, respectively from right to left, with common ε_a and U ($\varepsilon_a=-8$ and $U=16$). (c) Curves of $n_\uparrow(t)-n_\downarrow(t)$; solid lines represent those from the solution of (3.3) and broken lines the results given in Fig. 1(a).

From these results we see that the initial stage where $n_\uparrow(t)=n_\downarrow(t)$ is almost common in the three cases and the final stage where $n_\uparrow(t)\neq n_\downarrow(t)$ is also almost common. Separation in time between the final stages of the cases of $\varepsilon_m=0.001$ and 0.01 is the same as that between those of $\varepsilon_m=0.01$ and 0.1. Those behaviors of $n_\sigma(t)$ suggest that variation of $n_\uparrow(t)-n_\downarrow(t)$ shows an exponential growth. In fact those are explained by the following simple analysis.

By putting $u(t)=\vartheta(t)$ (the step function), $\varepsilon_a(t)=\varepsilon_a$ and $U(t)=U$, we have the following expression,

$$n_\sigma(t) = \frac{1}{2}[1-\exp(-2t)] + \frac{1}{\pi}\int_0^t d\tau \int_0^t d\tau' \,\frac{\exp(-2t+\tau+\tau')}{\tau-\tau'}$$
$$\times \sin\Big[\int_{\tau'}^{\tau} dt'[(U/2)\{1-2n_{-\sigma}(t')\}+\sigma\varepsilon_m]\Big]. \qquad (3.1)$$

Using the mean value approximation,

$$\sin\Big[\int_\tau^{\tau'} dt'[(U/2)\{1-2n_{-\sigma}(t)\}+\sigma\varepsilon_m]\Big] \simeq \sin\Big[(U/2)\{1-2n_{-\sigma}(\tfrac{\tau+\tau'}{2})\}(\tau-\tau')\Big], \qquad (3.2)$$

putting $n_\sigma(t)=n(t)+\sigma\varepsilon_m(m(t)/2)$ and taking only the linear term in ε_m, we obtain the differential equation,

$$d^2m(t)/dt^2+4dm(t)/dt+4[1-(U/\pi)]m(t)$$
$$= (4/\pi)[2-\exp(-t)-Um(t/2)\exp(-t)]. \qquad (3.3)$$

Neglecting the $m(t/2)$ term (justified by the obtained solution), we can solve (3.3) with the initial condition $m(0)=(dm/dt)_{t=0}=0$. The obtained expression of $m(t)$ for larger t is

$$m(t) = [4\pi\{1-(U/\pi)^{1/2}\}\{1-2(U/\pi)^{1/2}]^{-1}\exp[2\{(U/\pi)^{1/2}-1\}t]. \qquad (3.4)$$

This represents indeed an exponential growth of $m(t)$ for $U>\pi$ which is the condition for the stable magnetic solution in equilibrium. In Fig. 1(c) we show comparison between results from the solution of (3.3) and the numerical calculation.

4. Energy Dissipation

In this section we show first the numerical results on $n_\sigma(t)$ with the initial condition that $n_\uparrow(-\infty)=1$ and $n_\downarrow(-\infty)=0$ and then calculate energy dissipation. Here $u(t)$ is assumed to be $\exp[-(t/T)^2]$, and ε_a and U are taken to be constant as $\varepsilon_a=-1$ and U=2. T is assumed to be in the range of 9 to 11. This set of parameters and the initial condition enable us to use the available formula for the energy dissipation in the slow limit. $n_\sigma(t)$ are shown in Fig. 2(a) in the case of T=10. It is noted that since $u(t)$ has a form given above and $\varepsilon_a=-1$ and U=2 (the symmetric case), the instantaneous ratio, $U/\Delta(t)$, which is $U/u(t)$ in our unit, passes the boundary ($U/u(t)=\pi$) between the magnetic and nonmagnetic solutions in equilibrium. It is also noted that T is large and we have a slow case, but there is still asymmetry left in spin space due to the initial condition. The artificial magnetic field ε_m is unneccessary in this case.

The $P(\varepsilon)$ function, which describes energy dissipation, is defined by

$$P(\varepsilon) = \langle\infty|\delta(\varepsilon-H_\infty+E_0)|\infty\rangle, \qquad (4.1)$$

where $|\infty\rangle$ means the final state of the system in the Schrödinger

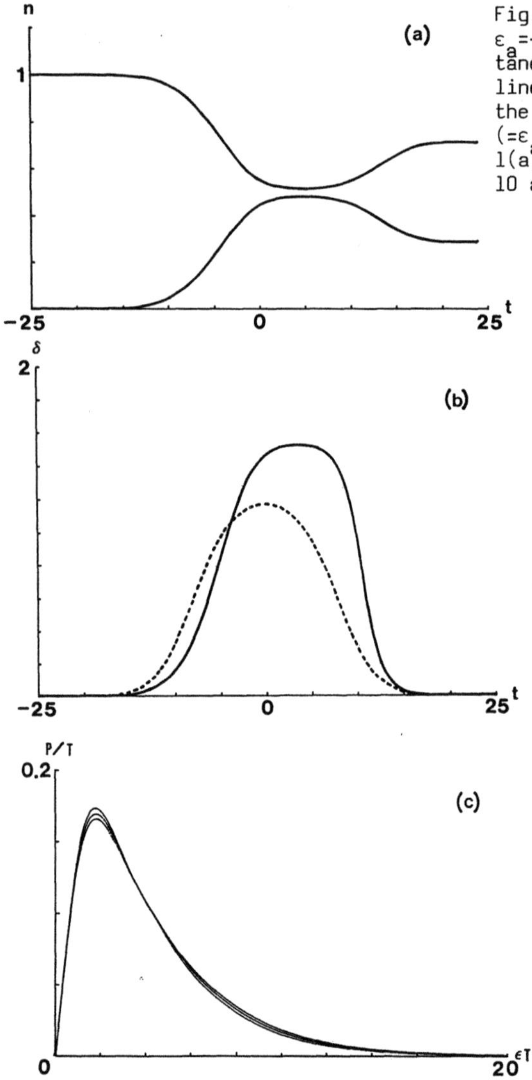

Fig. 2 (a) $n_\sigma(t)$ in the case of $\varepsilon_a=-1$, U=2 and T=10. (b) Instantaneous phase-shifts, the solid line represents $|\delta_\sigma(t)-(\pi/2)|$ and the broken line that with $\varepsilon_a=0.424$ ($=\varepsilon_a+Un_{-\sigma}(\infty)$ in the case of Fig. 1(a)). (c) $P(\varepsilon)/T$ vs. εT for T=9, 10 and 11.

representation, H_∞ denotes H given by (2.2) at $t=\infty$, and E_0 the energy of the initial state. The Fourier transform of $P(\varepsilon)$, $P(t)$ is given in terms of the instantaneous phase-shifts as

$$P(t) = \exp[\pi^{-2}\int_0^\infty d\omega\ \omega\Sigma_\sigma|\delta_\sigma(\omega)|^2\{\exp(-i\omega t)-1\}], \tag{4.2}$$

$$\delta_\sigma(\omega) = \int_{-\infty}^{\infty} dt\ \exp(i\omega t)\delta_\sigma(t), \tag{4.3}$$

where $\delta_\sigma(t)$ is the instantaneous phase-shift defined by

$$\delta_\sigma(t) = (\pi/2)-\tan^{-1}[(\varepsilon_a+Un_{-\sigma}(t))\exp\{(t/T)^2\}]. \quad (4.4)$$

Calculated results of $\delta_\sigma(t)-(\pi/2)$ from the obtained $n_\sigma(t)$ are shown in Fig. 2(b). together with a curve of $\delta_\sigma(t)-(\pi/2)$ of a constant ε_a and U=0 for comparison. Note that $|\delta_\uparrow(t)-(\pi/2)|=|\delta_\downarrow(t)-(\pi/2)|$.

Calculated results of $P(\varepsilon)$ are shown in Fig. 2(c) for T=9, 10 and 11. $P(\varepsilon)$ contain the delta function part not shown, which is given by

$$\delta(\varepsilon)\exp[-2\pi^{-2}\int_0^\infty d\omega\ \omega|\delta_\uparrow(\omega)|^2]. \quad (4.5)$$

The factor (the Debye-Waller factor) in (4.5) is calculated to be 0.1447, 0.1357, 0.1281 for T=9, 10, 11, respectively. It is to be noted that, if U=0, there is a scaling property for $P(\varepsilon)$. It is proved that $P(\varepsilon)/T$ vs. εT gives a universal curve for any T and the Debye-Waller factor does not depend on T if U=0 [7]. Because of the Hartree-Fock field this scaling property does not hold any more as is seen in Fig. 2(c).

5. Summary and Discussion

The effects of the intraatomic Coulomb interaction are examined in the time-dependent Hartree-Fock approximation in the time-dependent Newns-Anderson model. We found that some asymmetry in spin space is neccessary for the system to cross the boundary between the "nonmagnetic and magnetic" regions. We found also an exponential growth for the developement of the "magnetic" solution. On the energy dissipation,the scaling law, which is known for the no interaction case, is found not to be held because of the Hartree-Fock field.

The Hartree-Fock approximation has defects, which are known in this system in equilibrium; the magnetic-nonmagnetic sharp transition predicted in the approximation should not exist. In equilibrium we know now all physics on the Anderson model and how good (or bad) the Hartree-Fock aproximation is. Knowing this,we might guess some physics from the results by the time-dependent Hartree-Fock approximation. On the intraatomic Coulomb interaction effects on the charge transfer problem, the important quantity is the ratio r_c,

$$r_c = [\tfrac{1}{2}\Sigma_\sigma I_\sigma^0(\infty)+I^-(\infty)]^2/I^-(\infty), \quad (5.1)$$

where I_σ^0 and I^- are defined more generally as $I_\sigma^0(t)=n_\sigma(t)-\langle\hat{n}_{a\uparrow}(t)\hat{n}_{a\downarrow}(t)\rangle$ and $I^-(t)=\langle\hat{n}_{a\uparrow}(t)\hat{n}_{a\downarrow}(t)\rangle$ instead of (2.5) and (2.6). Here the metal surface is assumed to be paramagnetic. If U=0, r_c is unity. $r_c>1$ means that there is the effect of U. The present approximation tells $r_c=1$ when $n_\uparrow(t)=n_\downarrow(t)$ which is the nonmagnetic solution. $r_c>1$ occurs only for the magnetic solution ($n_\uparrow(t)\neq n_\downarrow(t)$). This explains partially the effect of the U term, but that is not entirely true. It is easy to see that the deviation $\langle\hat{n}_{a\uparrow}(t)\hat{n}_{a\downarrow}(t)\rangle-n_\sigma(t)^2$ occurs even in the first order pertubation of the U term. We emphasize here that to observe experimentally r_c it is very important to see the effect of the intraatomic Coulomb interaction.

In the present approximation we introduced ε_m. If physical quantities to calculate are insensitive to the magnitude of ε_m, namely to the time duration between the initial and final stages, the introduction of ε_m saves the difficulty. Otherwise, we have to know a way to estimate the magnitude of ε_m.

ε_m should represent an effect of the fluctuation, μ ($=\langle(n_{a\uparrow}(t)-n_{a\downarrow}(t))^2\rangle^{1/2}$). It is suggested one takes such a value of ε_m as producing the spin polarization corresponding to μ at an appropriate time, which is $[2n_\sigma(t)(1-n_\sigma(t))]^{1/2}$ in the nonmagnetic solution in the present approximation.

On the results of the energy dissipation, it is seen that, with increasing T, the Debye-Waller factor decreases in its magnitude, the width of $P(\varepsilon)/T$ increases and the peak of $P(\varepsilon)/T$ decreases. This is a tendency leaving the adiabatic limit in spite of increasing T, which means approaching the adiabatic limit. It is not clear at present why it is so, although the tendency is weak and the system, in total, is approaching the adiabatic limit.

Acknowledgments

The authors would like to thank K. Cho, H. Nagayoshi and other members of the group for discussion. One of the authors (A. Y.) would like to thank D. M. Newns of Imperial college, London for discussion during his stay in London and the Japan Society for Promotion of Science for making this visit possible.

References

1. A. Blandin, A. Nourtier and D. Hone, J. Phys. (Paris) 37, 369 (1976).
2. W. Bloss and D. Hone, Surface Sci. 72, 22 (1978).
3. J. K. Nørskov, D. M. Newns and B. I. Lundqvist, Surface Sci. 80, 179 (1979).
4. J. K. Nørskov and B. I. Lundqvist, Surface Sci. 89, 251 (1979).
5. G. Blaise and A. Nourtier, Surface Sci. 90, 495 (1979).
6. K. Schönhammer and O. Gunnarsson, Phys. Rev. B22, 1 (1980).
7. R. Brako and D. M. Newns, Solid State Commun. 33, 713 (1980).
8. K. Makoshi, J. Phys. C16, 3617 (1983).
9. K. Makoshi, H. Kawai and A. Yoshimori, J. Phys. Soc. Japan 53, 2441 (1984).

Ionization Probabilities of Desorbed Atoms

A. Okiji and N. Kawakami
Department of Applied Physics, Osaka University, Suita, Osaka 565, Japan

Ionization and neutralization probabilities of atoms desorbed from metal surfaces are calculated with the use of the Newns-Anderson model. The main purpose of this calculation is to examine the effect of the intraatomic Coulomb interaction on such probabilities. For the case where the desorption velocity v is infinite, the effect of the Coulomb interaction is treated exactly. For other cases of finite v, the retardation effect on the quantity is taken into account phenomenologically on the basis of a static solution of the model.

1. Introduction

The time-dependent Newns-Anderson model [1] has been widely employed in order to discuss the dynamical properties of atoms scattered (or sputtered) from metal surfaces. The microscopic calculation of the ionization probability on the basis of this model has been done firstly by BLANDIN et al.[2] and subsequently by several authors [3]. In these calculations, the intraatomic Coulomb interaction U in the outgoing atom is neglected because of its difficulty. Although an attempt to incorporate the U-term into the theory has been made by MAKOSHI et al. by adapting the Hartree-Fock approximation to the theory of BLANDIN et al.[4], their treatment leaves the many-body character of the U-term an open question. In this report, we make a simple estimation for the ionization probabilities of adatoms desorbed from metal surfaces, laying stress on the many-body effects of the Coulomb interaction on such probabilities. We consider the Newns-Anderson model described by the following Hamiltonian [1]:

$$H = \sum_{k,\sigma} \varepsilon_k n_{k\sigma} + \sum_{\sigma} \varepsilon_a(z) n_{a\sigma} + \sum_{k,\sigma} [V_k(z) C^+_{k\sigma} C_{a\sigma} + \text{H.C.}] + U(z) n_{a\uparrow} n_{a\downarrow} \quad , \tag{1.1}$$

where $C_{k\sigma}(C_{a\sigma})$ is the annihilation operator for an electron of the conduction band (adatom orbital), $n_{k\sigma}=C^+_{k\sigma}C_{k\sigma}$, $n_{a\sigma}=C^+_{a\sigma}C_{a\sigma}$ and z is a position of adatom measured from a metal surface (z=0 is assumed to be the equilibrium position of an adatom). The orbital energy $\varepsilon_a(z)$, the transfer matrix $V_k(z)$ and the intraatomic Coulomb interaction U(z) are considered to be time-dependent through the trajectory approximation z=z(t). Using this Hamiltonian, we discuss firstly the ionization probabilities for the case where the velocity of a desorbed atom is infinite, because in this limit we can make an exact calculation and the results are used in §3. The retardation effect caused by the time-dependent perturbation is taken into account phenomenologically in §3 and a brief summary is given in §4.

2. Case where the desorption velocity is infinite

For an adatom in the equilibrium state (t<0), the parameters ε_a, V_k and U are considered to be independent of time t. After the desorption occurs at t=0, we must take into account the time-dependent perturbation in the Hamiltonian (1.1). We can, however, simplify the problem in the case where the desorption velocity perpendicular to a surface, v, is very large. For instance, if we assume that $V_k(z)=V_k\exp(-az)$ and $z=vt$ $(t>0)$, the time-dependence of V_k can be written as $V_k(t)=V_k$ for $t<0$ and $V_k(t)=V_k\exp(-avt)$ for $t>0$ (ε_a and U are assumed to be time-independent). When v is large, $V_k(t)$ is seen to behave like a step function $V_k(t)\simeq V_k\theta(-t)$. Therefore in this limiting case the system reduces to one having the sudden-change potential. In this case we can calculate the ionization probabilities from the knowledge of the static solution of the Hamiltonian (1.1) for z=0 (t<0). The static solution of (1.1) has been already obtained exactly with the condition that the density of states for the conduction electrons is constant (wide band limit) and V_k does not depend on the wave vector k [5,6]. We adapt this solution to the system considered now, obtaining the ionization probabilities.

2-1 Zero temperature case

The static solution of the Hamiltonian (1.1) at zero temperature has been given in a rather simple form, which is described by a set of the linear integral equations,

$$\sigma_a(\Lambda)-\int_{-\infty}^{Q} R(\Lambda-\Lambda')\sigma_a(\Lambda')d\Lambda' = \frac{1}{4\pi}\int_{-\infty}^{\infty}\operatorname{sech}\pi(\Lambda-x^2)\frac{dx/\sqrt{u}}{(x+\sqrt{u}/2)^2+1/4u}\,, \tag{2.1}$$

$$\sigma_c(\Lambda)-\int_{-\infty}^{Q} R(\Lambda-\Lambda')\sigma_c(\Lambda')d\Lambda' = \frac{1}{4\pi}\int_{-\infty}^{\infty}\operatorname{sech}\pi(\Lambda-x^2)dx\,, \tag{2.2}$$

where R(x) is the function given by $[1+\exp(|\omega|)]^{-1}$ in Fourier space, $u=U/2\Delta$ and $E_a=\varepsilon_a/2\Delta$. Here, Δ is half of the resonance width of the adatom orbital ($\Delta=\pi\rho V^2$, ρ is the density of states for conduction electrons). The cut-off parameter Q is determined from the variational principle,

$$\int_{-\infty}^{Q}\sigma_c(\Lambda)d\Lambda = (u/2+E_a)/\pi\sqrt{u}\,, \tag{2.3}$$

where E_a is measured from the Fermi energy. The occupation number $n_a=\langle n_{a\uparrow}+n_{a\downarrow}\rangle$ and the binding energy E of the adatom can be described in terms of σ_a,

$$n_a = 1-\sqrt{u}\int_{-\infty}^{Q}\sigma_a(\Lambda)d\Lambda\,, \tag{2.4}$$

$$(E-E_s)/2\Delta = (u/2+E_a)n_a + 2u\int_{-\infty}^{Q}d\Lambda\sigma_a(\Lambda)\int_0^{\infty}dx\, x^2\operatorname{sech}\pi(\Lambda-x^2)\,, \tag{2.5}$$

where

$$(E_s-E_s(u{=}0))/2\Delta = \frac{1}{\pi}\mathrm{Re}(1+iu)\log(1+iu)-\int_0^{\infty}\frac{\sqrt{x}}{\pi}dx\int_0^{\infty}\frac{\operatorname{cosech}\pi(x+y^2)dy}{(y+\sqrt{u}/2)^2+1/4u}\,.$$

Here E_s is the binding energy for the symmetric case ($U=-2\varepsilon_a$). With the formulae (2.1)-(2.5), we can calculate the ionization probabilities directly; the probability for getting the negative ion is given by $P_-=\partial E/\partial U$, for the neutral atom by $P=n_a-2P_-$ and for the positive ion by $P_+=1-P_--P$. Although the analytical evaluation of these probabilities is difficult in general, it is possible in some limiting cases [6]. We first discuss these cases. The basic method used below is the standard Wiener-Hopf technique developed for the integral equations [7].

(a) Symmetric limit $(u+2E_a)\ll\sqrt{u}$: In this case it is seen from (2.3),(2.4) and (2.5) that $(u+2E_a)/\sqrt{u}$, n_a and $\partial E/\partial U$ can be expanded in power series of $\exp(\pi Q)$. By eliminating $\exp(\pi Q)$ from these series, we can get the expression for the ionization probabilities in series expansion form of $c=(u+2E_a)/2\sqrt{u}$:

$$P_- = P_-^s - \frac{\sqrt{u}}{2}\chi_1 c - \frac{u}{2}\frac{d\chi_1}{du}c^2 - \frac{\pi\sqrt{u}}{4}(\chi_1-\sqrt{3}\chi_3)c^3 + O(c^4) \quad ,$$

$$P = 1 - 2P_-^s + u\frac{d\chi_1}{du}c^2 + O(c^4) \quad , \tag{2.6}$$

$$P_+ = 1 - P_- - P \quad , \quad \chi_n = \frac{1}{\pi}\int_{-\infty}^{\infty} \frac{1/u}{(y+\sqrt{u}/2)^2+1/4u} e^{-\pi n y^2} dy \quad ,$$

where P_-^s is the ionization probabilities for the negative ion in the symmetric case, given by $P_-^s = 1/2+dE_s/dU$.

(b) Asymmetric limit $(u+2E_a)>>\sqrt{u},1$: In this case the configuration restriction retains the double occupation probability P_- to be zero. Furthermore, the effect of the Coulomb interaction on P and P_+ appears only through the expression of the renormalized orbital level [8] $\varepsilon_a^*=\varepsilon_a+(\Delta/\pi)\ln(\pi eu/2)$:$P=n_a(\varepsilon_a^*)$, $P_+=1-n_a(\varepsilon_a^*)$. The universal curve for $n_a(\varepsilon_a^*)$ is given in ref.[6]:

$$n_a(\varepsilon_a^*) = 1 - \frac{i}{\sqrt{2\pi}}\int_{-\infty}^{\infty} \frac{d\omega}{\omega+i0} \frac{\exp(-|\omega|-i\varepsilon_a\omega/2\Delta)}{G(\omega)} \quad , \tag{2.7}$$

where

$$G(\omega) = \sqrt{2\pi}\left(\frac{i\omega+0}{2\pi e}\right)^{i\omega/2\pi} / \Gamma\left(\frac{1}{2}+i\omega/2\pi\right) \quad .$$

Numerical results for the ionization probabilities are shown in Fig.1. In Fig.1-(a) the probabilities are shown as a function of ε_a. Since ε_a is measured from the Fermi energy in this calculation, the variation of ε_a can be also regarded as that of the work function. In the region where $(u+2E_a)$ is small, the results are well described by (2.6). Figure 1-(b) shows the probabilities as a function of U, choosing a typical value of ε_a being zero. It is seen that P_- can be regarded as zero in the suitably large-U region ($U/2\Delta \gtrsim 5$). In this region the probabilities for getting the positive ion and the neutral atom are given by the formula (2.7).

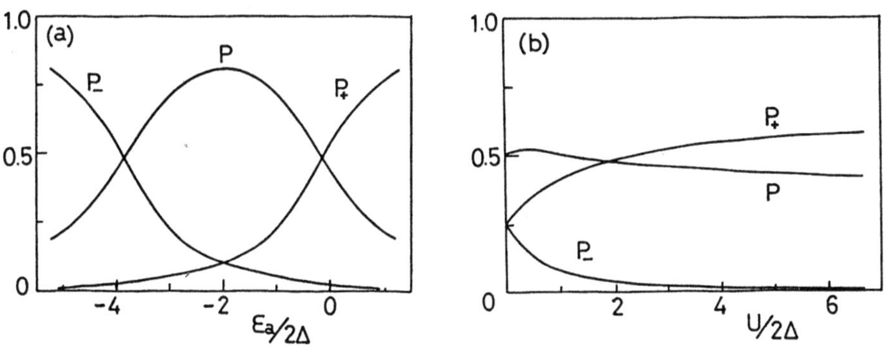

Fig.1. Ionization probabilities at zero temperature: (a)$U/2\Delta=4$, (b)$\varepsilon_a=0$

2-2 Finite temperature case

The static solution of the Hamiltonian (1.1) has been already extended to the finite temperature case [6]. Since the expression for the thermodynamic equations is rather complicated, we show only the numerical results for the ionization probabilities below.

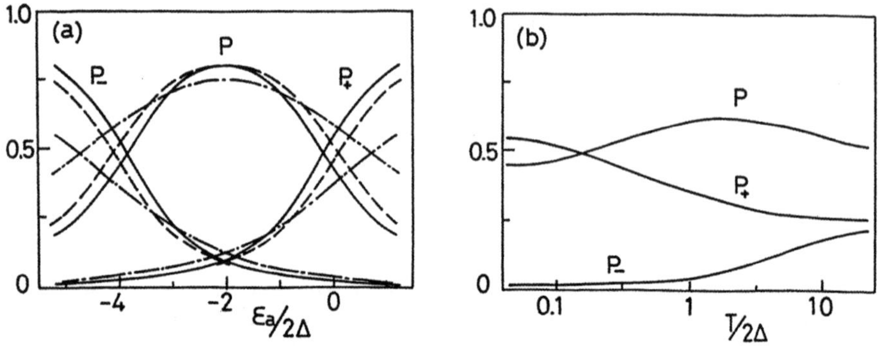

Fig.2. Ionization probabilities at finite temperatures: (a) $U/2\Delta=4$; $T/2\Delta=0$ ———, 0.32 – – – and 1.0 —·— , (b) $\varepsilon_a=0$ and $U/2\Delta=4$

In Fig.2-(a) the ionization probabilities are shown as a function of ε_a for given temperatures. It is seen that the probabilities become insensitive to the change of ε_a, as the temperature is increased. This is because thermal excitations may remove the configuration restriction gradually as the temperature is increased. Figure 2-(b) shows the temperature-dependence of the probabilities for the fixed values of $\varepsilon_a=0$ and $U/2\Delta=4$. In the high temperature region, $P_\pm$(P) takes the value of 1/4 (1/2), since all the configurations are equally thermally populated. As the temperature is decreased, thermal excitations to the configuration space of $n_a=2$ is gradually suppressed, so that P_- becomes very small and the probabilities for getting positive ions and neutral atoms become dominant in the case of ε_a being small and U being large.

3. An examination of the retardation effect

As discussed in §2, the ionization probabilities are dominated by the initial memory of the static state (t<0), when the desorption velocity v is very large. For arbitrary values of v, however, we must consider the retardation effect caused by the explicitly time-dependent parameters ε_a, Δ and U. Since the microscopic treatment is quite difficult for such a time-dependent Hamiltonian including the Coulomb interaction U, we take into account the retardation effect phenomenologically in this section.

First, we notice that we can get the static solution and the corresponding "static ionization probabilities" $P_\pm(z)$ exactly for a given position z of the adatom, by regarding the parameters ε_a, Δ and U appearing in §2 as $\varepsilon_a(z)$, $\Delta(z)$ and $U(z)$. We divide the time interval $0<t<\infty$ into the infinitesimal time intervals $dt_1, dt_2, \cdots$, that is, $t_0=0$, $t_1=t_0+dt_1$, $t_2=t_1+dt_2, \cdots$. Our phenomenological assumption to treat the retardation effect is as follows: in the small time interval dt_n, the "static ionization probabilities" $P_\pm(z)$ corresponding to $z=vt_{n-1}$ may decay at the rate of $(1-2\Delta(vt_n)dt_n)$ and $P_\pm(z)$ corresponding to $z=vt_n$ may come in at the rate of $2\Delta(vt_n)dt_n$. For instance, at the time $t_1=dt_1$ the ionization probabilities take the values of

$$[1-2\Delta(vt_1)dt_1]P_\pm(z=0) + 2\Delta(vt_1)dt_1P_\pm(z=vt_1) \quad . \tag{3.1}$$

Similarly, the expression for the ionization probabilities $\tilde{P}_\pm$ at $t\to\infty$, which we seek, can be written as

$$\begin{aligned}\tilde{P}_\pm &= P_\pm(0)\prod_{m=1}^{\infty}(1-2\Delta_m dt_m) + \sum_{\ell=1}^{\infty} P_\pm(vt_\ell)2\Delta_\ell dt_\ell \prod_{m=\ell+1}^{\infty}(1-2\Delta_m dt_m) \\ &= P_\pm(0)\exp\left[-2\int_0^\infty \Delta(z)\frac{dz}{v}\right] + \int_0^\infty 2\Delta(z)P_\pm(z)\frac{dz}{v}\exp\left[-2\int_z^\infty \Delta(z')\frac{dz'}{v}\right] \quad ,\end{aligned} \tag{3.2}$$

where $P_\pm(z)$ is the "static ionization probability" at the given position z, which can be calculated exactly as mentioned earlier in this section. Noting that $P_+(z)+P_-(z)+P(z)=1$, we can easily show that the relation $\tilde{P}_+ + \tilde{P}_- + \tilde{P} = 1$ holds. When $v\to\infty$, $\tilde{P}_\pm$ is equal to $P_\pm(z=0)$, which is evaluated in §2, and when v=0, $\tilde{P}_\pm$ reproduces the adiabatic results.

In order to see above phenomenological treatment more clearly, we consider a simple scattering problem for the case; $\Delta(t)=\Delta_0$ for $0<t<t_0$ and $\Delta(t)=0$ for other times; ε_a and U are time-independent constants. In this case the ionization probabilities at $t\to\infty$ are calculated as

$$\tilde{P}_\pm = P_\pm(\Delta=\Delta_0) + [P_\pm(\Delta=0)-P_\pm(\Delta=\Delta_0)]\exp(-2\Delta_0 t_0) \quad , \tag{3.3}$$

which, we see, includes a type of retardation effect for the sudden change potential. In the symmetric case of U=0, which means that ε_a equals to the Fermi energy, this expression coincides with the result obtained from the theory of BLANDIN et al.[2].

For a typical example of $\Delta(z)=\Delta_0\exp(-az)$, (3.2) is rewritten as

$$\tilde{P}_\pm = P_\pm(\Delta=\Delta_0)\exp(-\frac{2\Delta_0}{va}) + \int_0^{\Delta_0} P_\pm(\Delta(z))\exp(-\frac{2\Delta}{va})\frac{2d\Delta}{va} \quad . \tag{3.4}$$

when v is large (fast case), (3.4) is reduced to

$$\tilde{P}_\pm \simeq P_\pm(\Delta=\Delta_0) - \frac{2\Delta_0}{va}[P_\pm(\Delta_0) - \frac{1}{\Delta_0}\int_0^{\Delta_0} P_\pm(\Delta)d\Delta] \quad . \tag{3.5}$$

Therefore for large v, the correction of the ionization probabilities can be expressed by the term proportional to 1/v, which is considered to be a plausible result [2]. On the other hand, when v is small, (3.4) is reduced to

$$\tilde{P}_\pm \simeq P_\pm(\Delta=0) + (ab_1/2)v \quad , \tag{3.6}$$

where we assumed that $P_\pm$ can be expanded in the following form: $P_\pm=P_\pm(\Delta=0)+b_1\Delta+\cdots$. The correction term expressed by $(ab_1/2)v$ differs from that of $\exp(-c/v)$ which is expected by earlier treatments [2,3]. This difference means that our phenomenological treatment may overestimate the retardation effect in the small v region. In Fig.3 the ionization probabilities $\tilde{P}_\pm$ given by (3.4) are plotted as a function of (1/v) for the symmetric case $U=-2\varepsilon_a$ (U and ε_a are assumed to be time-independent). In this numerical calculation, the exact solution of the static problem is used for the evaluation of $P_\pm(z)$ in (3.4). Note that $\tilde{P}_\pm=1/4$, $\tilde{P}=1/2$ and these values are irrespective of v when U=0.

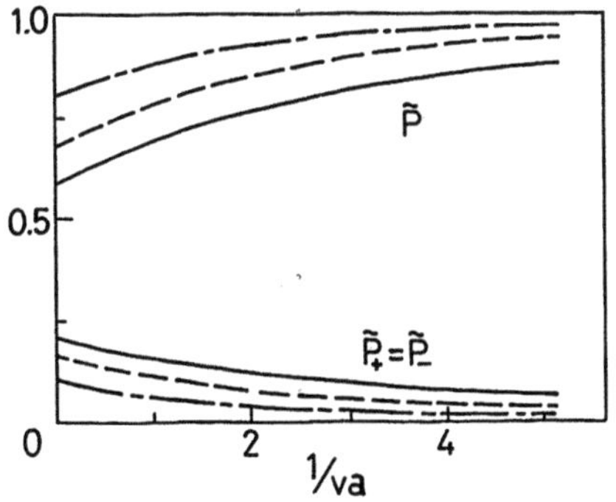

Fig.3 Ionization probabilities given by (3.4) for the symmetric case U=1.0 ——— , 2.0 – – – , and 4.0 —·— (Δ_0=0.5)

4. Summary

We have estimated the ionization probabilities of atoms desorbed from metal surfaces with the use of Newns-Anderson model, laying stress on the effect of the intraatomic Coulomb interaction U. The effect of U is taken into account exactly for the case where the desorption velocity v is infinite. As for the case of finite v, a simple phenomenological treatment of the retardation effect is presented on the basis of the (static) exact solution. Since this phenomenological discussion is rather rough, that is, the effect of U is taken into account only through the "static ionization probabilities" $P_{\pm}(z)$, more precise treatment of retardation effect should be required to understand many-body aspects in such dynamical processes. Nevertheless, the estimation made here might be a starting point of understanding such an interesting phenomenon.

References

1. P.W. Anderson: Phys. Rev. 124, 41 (1961)
   D.M. Newns: Phys. Rev. 178, 1123 (1969)
2. A. Blandin, A. Nourtier and D.W. Hone: J. Physique 37, 369 (1976)
3. R. Brako and D.M. Newns: Surface Sci. 108, 253 (1981)
   T.B. Grimley, V.C. Jyothi Bhasu and K.L. Sebastian, Surface Sci. 124, 305 (1983)
4. K. Makoshi, H. Kawai and A. Yoshimori: preprint
5. P.B. Wiegmann: Phys. Lett. 80A, 163 (1980)
   N. Kawakami and A. Okiji: Phys. Lett. 86A, 483 (1981)
6. A. Okiji and N. Kawakami: Phys. Rev. Lett. 50, 1157 (1983)
   N. Kawakami and A. Okiji: Phys. Rev. Lett. 51, 2011 (1983)
   A. Okiji and N. Kawakami: J. Appl. Phys. 55, 1931 (1984)
   P.B. Wiegmann and A.M. Tsvelick: J. Phys. C16, 2281 (1983)
7. P.M. Morse and H. Feshbach: Methods of Theoretical Physics (McGraw-Hill, New York 1953)
8. F.D.M. Haldane: Phys. Rev. Lett. 40, 416 (1978)

Density-Functional Studies of the Atom-Surface Interaction and the Ionization Probability of Sputtered Atoms

N.D. Lang
IBM Thomas J. Watson Research Center, Yorktown Heights, NY 10598, USA

Abstract

A brief description of studies of the atom-surface interaction using density-functional methods is given. The ionization probability of atoms sputtered from metal surfaces is then discussed using results of these studies in conjunction with a resonant tunnelling approach. Good agreement of this theory with experimental data of Yu on the sputtering of adsorbed alkali atoms is obtained.

We begin our discussion with a brief outline of the calculation of the ground-state properties of an atom interacting with a metal surface a fixed distance away. [1] We simplify the description of the metal substrate by using the jellium model to represent it: that is, the semi-infinite lattice of positive ions of the metal is smeared out into a semi-infinite homogeneous positive background, This model eliminates much of the detail of the substrate (which is now specified only by the value of its average electron density), but it is able to represent many of the properties of the metal that will interest us here. [1]

Our study of the metal-atom system is based on the density-functional theory of inhomogeneous electron systems due to HOHENBERG, KOHN and SHAM. [2] The equations that must be solved self-consistently to obtain the electron density distribution have a Hartree-like form, with a potential that includes not only electrostatic but exchange and correlation effects as well. For the latter we make the commonly used local-density approximation: the relevant exchange and correlation properties in any volume element are taken to be those of a uniform infinite electron gas with a density equal to that in the volume element. [2]

When an atom interacts with a system whose electronic states form a continuum, the discrete levels of the atom which are degenerate with the continuum broaden into resonances. The resonances formed in this way when Li, Si and Cl atoms interact with a high-electron-density metal are shown in Fig. 1 (at their calculated equilibrium separations). [3] (Explicitly, what is shown is the difference in density of eigenstates between the metal-atom system and the bare metal.) The states constituting the Cl 3p resonance are below the Fermi level,

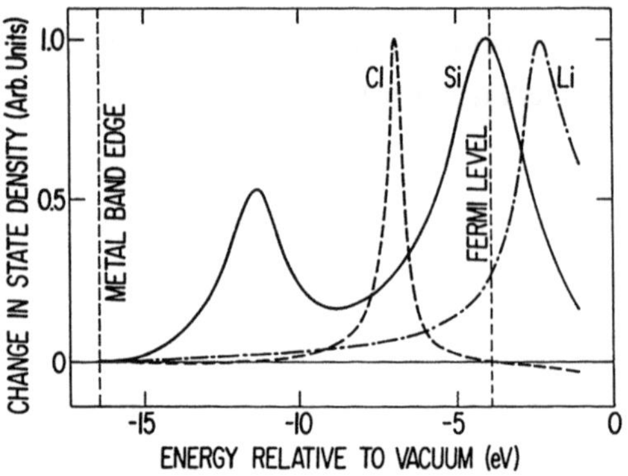

Fig. 1. Change in eigenstate density due to atomic chemisorption. The curves correspond to metal-atom separations which minimize the total energy. The lower Si resonance corresponds to the 3s level; the corresponding level for Cl is a discrete state below the bottom of the metal band. The metal electron density is 0.03 electrons/bohr^3. (From Ref. 3.)

and are therefore occupied; those constituting the Li 2s resonance are essentially empty. This implies that charge transfer has taken place, toward the Cl and away from the Li, as would be expected from the electronegativities of these atoms. The prohibitively large energy required either to fill or empty the Si 3p level forces this resonance to straddle the Fermi level, resulting in the formation of a covalent, rather than ionic, bond.

The electron densities [3] associated with the three fundamental bond types are exhibited for comparison in Fig. 2. In the top row of the figure are presented contours of constant total electron density for the three prototype systems. Note the way in which the contours rapidly regain their bare-metal form away from the immediate region of the atom. This is just a manifestation of the short range of metallic screening. Note also the way in which the metal contours bend toward the positively charged Li and away from the negatively charged Cl.

The detailed charge rearrangements associated with the atom-surface interaction are displayed in the second row of Fig. 2, which gives contour maps of the difference between the electron density in the metal-atom system and the superposition of bare-metal and free-atom densities. The solid contours indicate regions of charge accumulation; broken contours indicate regions of charge depletion.

The density-difference contours in the case of Li reveal the complexity underlying the notion of charge transfer. Taken alone, the density-difference plot indicates only that electrons have been displaced from the vacuum side to the metal side of the atom. The plot does not distinguish between the two possible interpretations of the displacement: polarization of the atom, or ionization

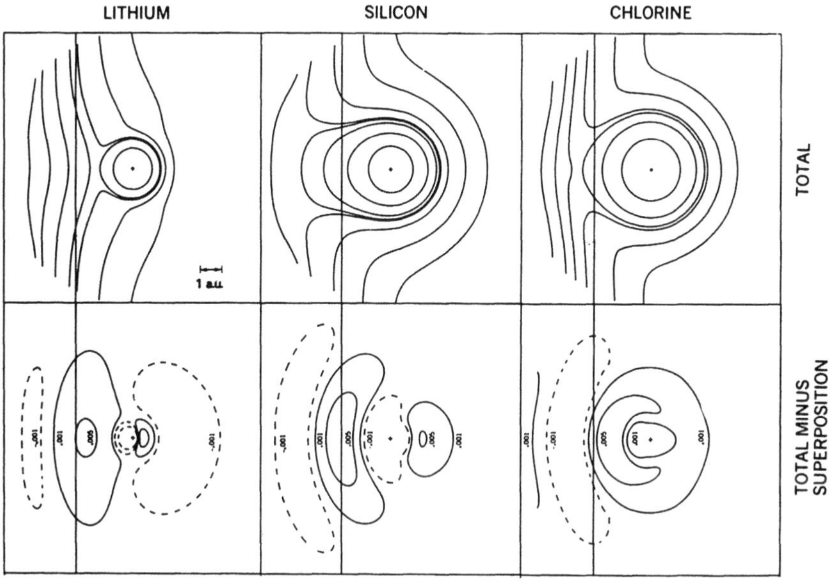

Fig. 2. Upper row: contours of constant electron density in a plane normal to the metal surface containing the atom nucleus (indicated by +): the metal is to the left-hand side; the solid vertical line indicates the positive-background edge. Atoms are at equilibrium distance from the metal. Contours are not shown outside inscribed circle of each square; contour values are selected to be visually informative. Second row: total electron density minus superposition of atomic and bare-metal electron densities (electrons/bohr^3). (From Ref. 3.)

(followed by metallic screening). The state-density plot in Fig. 1, however, resolves the ambiguity; the fact that virtually none of the states in the resonance associated with the Li 2s level are occupied clearly indicates that the ionization/screening interpretation is the correct one.

We now consider the ionization probability of secondary atoms emitted from layers chemisorbed on a metal surface. We use the results of our density-functional calculations to determine the values of the parameters (such as resonance width at equilibrium) that appear in our analysis; this analysis follows the approach discussed by BLANDIN, NOURTIER and HONE [4], and later by NØRSKOV and LUNDQVIST [5], BRAKO and NEWNS [6], and LANG. [7] The metal is treated as a non-interacting Fermi gas of work function Φ. We consider the interaction of the metal with a single atom, which moves along a classical trajectory $\vec{r}(t)$. The atom has a non-degenerate valence state $|a\rangle$ of energy ε_a, lying in the metal conduction band. This level has a finite width (as in Fig. 1) because of its interaction with the states $|k\rangle$ of the metal, via matrix elements $V_{ak} = \langle a|V|k\rangle$, with V the perturbation due to the atom-metal interaction. Charge transfer between atom and metal is taken to involve the transfer of an electron between the broadened level $|a\rangle$ and metal states of the same energy. (We thus assume that Auger processes can be neglected relative to this resonant transfer process. [8])

Use of the trajectory approximation means that we can write ε_a and V_{ak} in the electronic Hamiltonian as $\varepsilon_a(\vec{r}(t))$ and $V_{ak}(\vec{r}(t))$, with $\vec{r}(t)$ the trajectory. Spin effects are neglected. The Hamiltonian is then written [9]

$$H(t) = \sum_k \varepsilon_k n_k + \varepsilon_a(\vec{r}(t))n_a + \sum_k [V_{ak}(\vec{r}(t))c_a^\dagger c_k + \text{h.c.}], \tag{1}$$

where $c_a^\dagger$ and $c_k^\dagger$ are electron creation operators for $|a\rangle$ and $|k\rangle$ respectively, and $n = c^\dagger c$. We do not take account of interactions of the form $Un_{a\uparrow}n_{a\downarrow}$ in H, which would represent a major complication in the theory. We wish instead to discuss the simplest possible model that gives a reasonable picture of the experimental data.

Following Ref. 4 the simplifying assumptions are made that the $\vec{k}$ and $\vec{r}$ dependence of $V_{ak}(\vec{r}(t))$ are separable:

$$V_{ak}(\vec{r}(t)) = V_k u(\vec{r}(t)) \tag{2}$$

and that the energy-dependence of the level-width function

$$\Delta(\varepsilon) = \pi \sum_k |V_{ak}|^2 \delta(\varepsilon - \varepsilon_k) \tag{3}$$

can be neglected. The instantaneous resonance width is then given by

$$\Delta(\vec{r}(t)) = \Delta_0 |u(\vec{r}(t))|^2 \tag{4}$$

where we take $|u|^2 = 1$ for $t = 0$. Note that Δ is the half width at half maximum of the adsorbate resonance. With these assumptions, an expression for the occupation probability $\langle n_a(t)\rangle$ has been obtained in Refs. 4-7.

We neglect any dependence parallel to the surface, and take Δ and ε_a to depend only on z, the coordinate along the surface normal. We set $\hbar = 1$. We also use a simple exponential form for Δ:

$$\Delta(z) = \Delta_0 e^{-\gamma z} \tag{5}$$

($z = 0$ is the atom position at $t = 0$).

We will give the occupation probability in the limit $t \to \infty$ for a simple case in which ε_a varies linearly with distance and in which the velocity is constant (see Ref. 7). In particular, we take

$$\varepsilon_a(z) = \varepsilon_F + b(z - z_c) \tag{6}$$

with z_c the distance at which the level crosses ε_F, and we take $z = v_\perp t$. Neglecting terms of $O(e^{-\frac{\Delta_0}{\gamma v_\perp}})$, which is generally a good approximation [7],

$$<n_a(\infty)> = \frac{1}{2} + \frac{\zeta}{\pi}\mathrm{Re}\int_0^\infty \frac{dx}{\cosh x}\left(\frac{2\Delta_0}{\gamma v_\perp}\cosh x\right)^{i\zeta x} e^{-i\zeta\gamma z_c x}\,\Gamma(-i\zeta x) \tag{7}$$

where

$$\zeta = \frac{2b}{\gamma^2 v_\perp} \ . \tag{8}$$

In the limit in which $\zeta \to -\infty$ (this sign of ζ corresponds to $b<0$, which as we shall see is appropriate for the case of positive-ion formation discussed here), the probability that the sputtered atom is a positive ion far from the surface is* [7]

$$P^+ \equiv 1 - <n_a(\infty)> = \exp\left(-\frac{2\Delta(z_c)}{\gamma v_\perp}\right). \tag{9}$$

Comparisons with evaluations of (7) show this limiting form to be quite accurate for many cases of interest.

We now consider the case of a sputtered adsorbed Cs atom, which was studied experimentally by YU. [11] At small coverages, Cs atoms are chemisorbed as Cs^+ on metal surfaces, with the empty 6s level lying above the metal Fermi level just as the empty Li 2s level lies above the Fermi level in Fig. 1. Hence if the work function Φ of the surface is larger than the ionization potential I of Cs (3.9 eV), as indeed it is in Yu's experiment, the 6s level of the sputtered Cs atom will always face empty states of the metal as the Cs atom escapes and little neutralization by electron tunneling can occur. By using an adatom dipole layer to adjust the relative positions of the Fermi level and the vacuum level (this is done in the experiment by using adsorbed Li to decrease Φ, beyond the amount which the Cs itself decreases it), the 6s level can be deliberately forced to "cross" the Fermi level (when $I > \Phi$) as the Cs atom escapes, making electron tunneling energetically possible.

When the sputtered alkali atom is far from the surface, the valence level is an energy I below the vacuum level: $\varepsilon_a(\infty) - \varepsilon_F = \Phi - I$. For atom positions closer to the surface, the level $\varepsilon_a(z)$ is higher because of the image effect, i.e. it is easier to remove an electron from the level because the final-state energy of the system is lowered by the image interaction between the alkali and the metal. We thus write (in atomic units)

---

* This result has a simple semi-classical derivation (see e.g. Ref. 10). It was also obtained earlier by BRAKO and NEWNS (Ref. 6), although with a different criterion of validity.

$$\varepsilon_a(z) - \varepsilon_F = \Phi - I + \frac{1}{4(z - z_{im})} , \tag{10}$$

where z_{im} is the position of the image plane. [12]

We will describe this behavior of ε_a by a linear form (6) in the vicinity of z_c, with a negative slope b. We employ the convention that the case in which there is no crossing corresponds to $z_c \to \infty$; and we take the velocity to be constant, which is appropriate for the sputtered-atom kinetic energy in the experiment (cf. Ref. 7). We then use expression (9) to determine the ionization probability. Note the central role in this expression of $\Delta(z_c)$, the resonance half-width at the Fermi-level crossing.

We see from (9) that P^+ does not begin to drop significantly as soon as Φ is decreased below I, i.e. as soon as there is a crossing of the curve $\varepsilon_a(z)$ by the Fermi level at some finite distance z_c. The reason is simply that Φ must continue to decrease until z_c comes in close enough to the metal to give an appreciable width $\Delta(z_c)$; before this happens, $P^+ \doteq 1$. As z_c then decreases further with Φ, P^+ decreases rapidly because of the rapid increase in $\Delta(z_c)$.

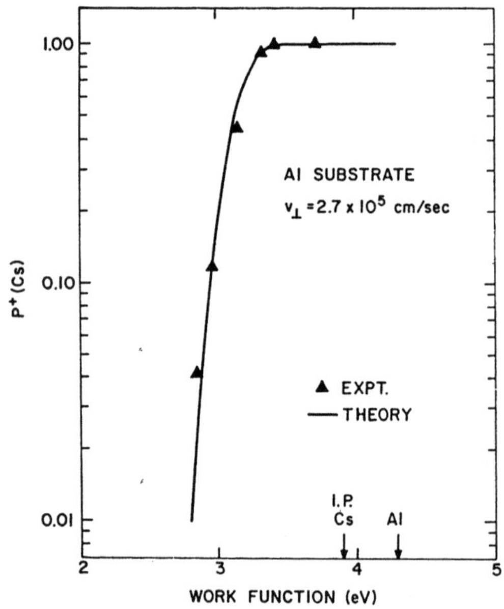

Fig. 3. Comparison between theoretical P^+ for Cs on an Al substrate and the corresponding experimental data. Given the approximate nature of the determination of such quantities as Δ_0 and γ, the extremely close agreement between theory and experiment should be regarded as somewhat coincidental; however it is clear that a reasonable account of the data is obtained from a completely *a priori* calculation. (From Ref. 11.)

In order to use (9), (5), and (10), which can be combined as

$$P^+ = \begin{cases} 1 & \Phi \geq I \\ & , \\ \exp\left(-A \exp\left[-\frac{\gamma}{4(I-\Phi)}\right]\right) & \Phi < I \end{cases} \quad (11)$$

where $A = 2\Delta_0(\gamma v_\perp)^{-1} \exp(-\gamma z_{im})$, to determine P^+ for the Cs/Al case, we must know the values of the quantities Δ_0, γ and z_{im}. Density-functional calculations for Cs of the type described above [3] as well as for the bare surface [12] give $\Delta_0 \sim 0.5$ eV and $z_{im} \sim -3$ bohr (recall that z is measured relative to the atom equilibrium position, with $z \rightarrow \infty$ in the vacuum) (see Ref. 11 for additional details). We furthermore use a value $\gamma = 0.7$ bohr^{-1}, determined as described in Ref. 11. Figure 3 compares the computed curve of P^+ vs. Φ with that determined experimentally by YU [11] for the Cs/Al case. The results are seen to be quite close.

References

1. For a general survey, see N. D. Lang: *Theory of the Inhomogeneous Electron Gas)*, ed. S. Lundqvist and N. H. March (Plenum, New York, 1983), pp. 309-389.
2. P. Hohenberg and W. Kohn: Phys. Rev. **136**, B864 (1964); W. Kohn and L. J. Sham: Phys. Rev. **140**, A1133 (1965).
3. N. D. Lang and A. R. Williams: Phys. Rev. Lett. **37**, 212 (1976); Phys. Rev. B **18**, 616 (1978).
4. A. Blandin, A. Nourtier, and D. W. Hone: J. Phys. (Paris) **37**, 369 (1976).
5. J. K. Nørskov and B. I. Lundqvist: Phys. Rev. B **19**, 5661 (1979).
6. R. Brako and D. M. Newns: Surf. Sci. **108**, 253 (1981).
7. N. D. Lang: Phys. Rev. B **27**, 2019 (1983).
8. J. K. Nørskov, D. M. Newns, and B. I. Lundqvist: Surf. Sci. **80**, 179 (1979).
9. P. W. Anderson: Phys. Rev. **124**, 41 (1961); D. M. Newns: Phys. Rev. **178**, 1123 (1969).
10. N. D. Lang and J. K. Nørskov: Physica Scripta **T6**, 15 (1983).
11. M. L. Yu and N. D. Lang: Phys. Rev. Lett. **50**, 127 (1983).
12. N. D. Lang and W. Kohn: Phys. Rev. B **7**, 3541 (1973).

Adsorbate-Surface Interactions

J.K. Nørskov

NORDITA, Blegdamsvej 17, DK-2100 Copenhagen, Denmark

Calculations of the adiabatic potential energy surface for atoms and molecules interacting with metal surfaces based on the effective medium theory are reviewed. For atomic chemisorption, the full potential energy surface has been calculated in a number of cases and the properties of the interaction potential can be related to the parameters describing the atom and surface in question. For molecular chemisorption and surface reactions, the effective medium theory can be used to make comparisons between binding energies and activation energies.

1. INTRODUCTION

The interaction of an atom or molecule with a metal surface is described to a large extent by the ground state adiabatic potential energy surface, which is the total ground state energy of the interacting system minus that of the constituents, calculated as a function of the coordinates of the adsorbates. In the adiabatic approximation, this is the potential energy surface on which the adsorbate nuclei move [1,2]. It determines equilibrium positions, chemisorption energies, vibrational frequencies, activation energies and, more generally, the dynamics of surface reactions. Even in cases where electronic excitations during a reaction are important, the adiabatic potential energy surface may serve as a starting point for a treatment of the dynamics [1,2].

Unfortunately, the combination of a semi-infinite solid and the low symmetry that the presence of the adsorbate usually imposes on the system makes a direct calculation of the potential energy surface very demanding, even with efficient methods like the local density approximation, to treat exchange and correlation effects [3]. Consequently, only a few calculations exist at present. One approach has been to approximate the surface by a (small) finite cluster of atoms [4]. Alternatively, a (thin) slab has been used [5,6]. If high coverages of adsorbates are considered, the translational symmetry along the surface can be exploited and the number of atoms per unit cell be kept low. Finally, perhaps the largest number of total energy calculations up to now have used a jellium model to describe the metal [7-9]. In this model, the metal ion cores have been smeared out as a homogeneous positive background, which is abruptly truncated at the surface.

An alternative approach is to make suitable approximations in the total energy calculation to enable a treatment of these low symmetry systems. The present paper is an overview of the insight which has been gained using one such method, the effective medium theory [10,11]. In particular, an attempt will be made to provide a conceptual picture of the interactions involved in chemisorption and to point out the important substrate and adsorbate parameters determining the properties of the interaction potential.

In the following, the effective medium theory is first briefly described. Then, in sect. 3, the properties of potential energy surfaces for atomic adsorption are discussed, using hydrogen and oxygen adsorption on transition metal surfaces as examples. Finally, in sect. 4, some properties of potential energy surfaces for molecular adsorption are discussed and the implications for surface reactions and catalysis are pointed out.

2. THE EFFECTIVE MEDIUM APPROACH

The effective medium approach is a perturbation theory which can be used to evaluate the difference δE in the binding energy of an atom (or molecule) when moved from one host to another. Treating the difference between the new and old host to first order in the region 'a' close to the atom where the atomic potential dominates, and the difference between the host with and without the atom to first order in the region outside, one gets [11]

$$\delta E = \int_a \delta\phi_0^{-a}(\vec{r})\Delta n(\vec{r})d\vec{r} + \delta\left\{\int_{-\infty}^{\varepsilon_F} \varepsilon\Delta n(\varepsilon)d\varepsilon\right\} . \tag{1}$$

Here the first, purely electrostatic term involves the difference $\delta\phi_0^{-a}(\vec{r})$ in electrostatic potential between the new and old hosts and the atom-induced charge density $\Delta n(\vec{r})$ in the old host. The integral is over the atomic region 'a' only and the superscript - a on $\delta\phi_0$ signifies that only the charge outside 'a' must be included. In the second term $\Delta n(\varepsilon)$ is the atom-induced density of states and the term involves the difference in the sum of atom-induced one-electron energies evaluated for a rigidly shifted potential inside region 'a'. The result (1) is derived within the local density approximation, but a similar result can be obtained for a non-local exchange-correlation energy. Both the derivation and the result closely resemble the force calculations of Andersen et al. [12].

The result (1) can be used in two different ways. If the energy of, say, an atom outside a metal surface is wanted, and it is possible to find a (high symmetry) system resembling the surface in the vicinity of the atom in which the interaction energy is known, then (1) can be used to give the absolute interaction energy outside the surface. One very convenient choice of starting point is a homogeneous electron gas. If the electron gas density is chosen as an average $\bar{n}_0$ of the true surface charge density $n_0(\vec{r})$ over the atom-induced electrostatic potential, then the atom-surface interaction energy ΔE can be written [11]

$$\Delta E(\vec{R}) = \Delta E_{\rm eff}^{\rm hom}(\bar{n}_0(\vec{R})) + \Delta E_{\rm cov}(\vec{R}) . \tag{2}$$

Here $\Delta E_{\rm eff}^{\rm hom}(\bar{n}_0)$ is the energy of the atom in a homogeneous electron gas of density $\bar{n}_0$ including the electrostatic term in (1), which can be shown to depend on $\bar{n}_0$ only [11]. The second term in (2) is just the one-electron energy term in (1).

An atom embedded in a homogeneous electron gas has spherical symmetry and a number of groups have calculated the binding energies [13]. Furthermore, the function $\Delta E_{\rm eff}^{\rm hom}(\bar{n}_0)$ need only be calculated once and for all. It is a property of the atom in question. Examples are shown in fig.1.

For a rare gas like He, the interaction is always repulsive. It is dominated by the kinetic energy cost on orthogonalizing more and more electron

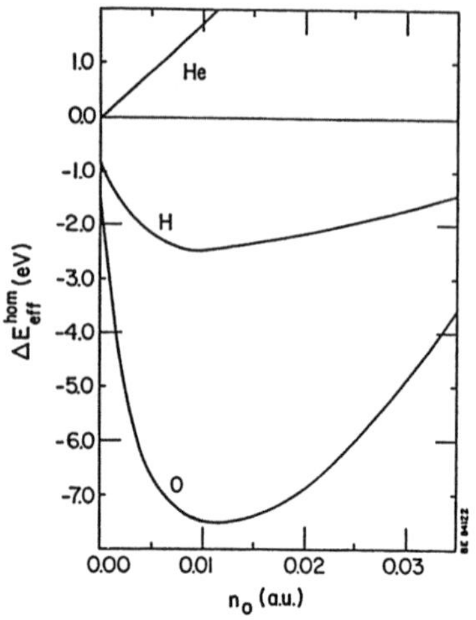

Fig.1. Interaction energy $\Delta E_{eff}^{hom}(n_0)$ of He, H and O with a homogeneous electron gas as a function of electron gas density n_0. From ref.[13]

gas states to the atomic states. For reactive gases like H and O, on the other hand, a minimum exists, indicating the willingness of these atoms to form chemical bonds. The low density limit of the interaction energy is -A, where A is the atomic affinity. The decrease in energy at densities larger than zero can be viewed as being due to the screening of the electron-electron repulsion in the negative ion. At even higher densities, the kinetic energy repulsion takes over, as for the rare gases.

Apart from the function $\Delta E_{eff}^{hom}(\bar{n}_0)$, one needs the surface charge density $n_0(\vec{r})$ to calculate the first term in (2). Self-consistent calculations of surface electron densities do exist now, but fortunately a mere superposition of atomic densities is usually an excellent approximation for the present purposes [14].

ΔE_{cov} reflects the difference in the one-electron spectrum on going from the homogeneous electron gas to the metal surface. If it is a free-electron-like metal, these changes are expected to be small. To first order in $(n_0(\vec{r}) - \bar{n}_0)$ one gets [11]

$$\Delta E_{cov}^{(1)}(\vec{R}) = \int_a (n_0(\vec{r}) - \bar{n}_0)\Delta V(|\vec{r} - \vec{R}|)d\vec{r} \ , \tag{3}$$

where ΔV is the atom-induced effective potential in the homogeneous electron gas. For transition metals, the interaction with the d-bands must be treated to higher order in the adsorbate-metal d interaction V_{ad}. The starting point is, however, not the free atom, but a renormalized atom embedded in an electron gas. As discussed above, such atoms can be regarded as screened negative ions and they have filled valence states well below the Fermi energy. The renormalized atom thus has a 'closed shell' well below the metal d-bands and the interaction is fairly weak. It is usually calculated using the resonant level model [15,16] taking parameters describing the d-bands and the hopping matrix element from ref. [12].

Another use of (1) is for estimating differences in interaction energies in cases where it is not possible to calculate absolute energies, or where they are not interesting. In the following two sections, examples of the two types of application of (1) will be given.

3. ATOMIC CHEMISORPTION

For atomic chemisorption the absolute interaction energies and their variation with the atomic positions can be well described starting from the atoms embedded in a homogeneous electron gas via (2). In the present section, properties of the atomic potential energy surface like the chemisorption energy, the bond length, the vibrational excitation energies and the activation energies for diffusion will be treated separately, and the main atom and surface parameters determining them will be discussed.

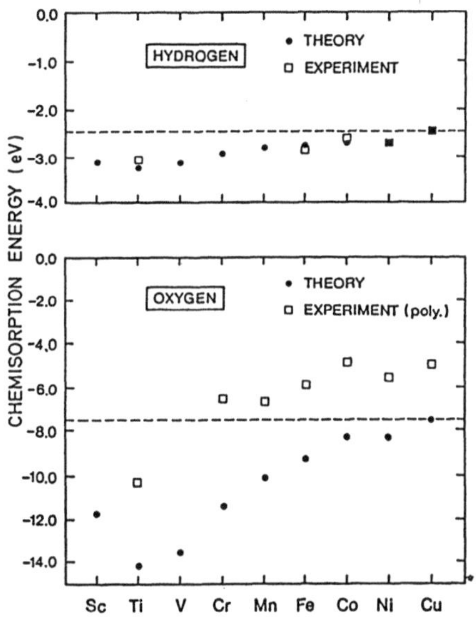

Fig.2. Comparison between calculated and experimental chemisorption energies for hydrogen and oxygen on the most close-packed surface of each of the 3d transition metals. The horizontal dashed line indicates the contribution to the chemisorption energy from $\Delta E^{hom}_{eff}(\bar{n}_0)$ in (2). This contribution is the same for all of the metals, since the optimum surface electron density where $\Delta E^{hom}_{eff}(\bar{n}_0)$ has its minimum value can always be found outside a surface. The trends along the series are given by ΔE_{cov} (see (2)). In the calculated values, no correction for the zero-point energy has been included. From refs.[15,17]

3.1. The chemisorption energy

In fig.2 the calculated chemisorption energies from (2) for hydrogen and oxygen on all the 3d-transition metals are compared to experimental values where found [15,17]. It is clear that the absolute magnitude as well as the trends are well described. Outside a surface, the optimum value of the density where $\Delta E^{hom}_{eff}(\bar{n}_0)$ has its minimum can always be found. The ΔE^{hom}_{eff} term in (2) thus always contributes the same amount to the chemisorption energy. The trends are therefore governed by ΔE_{cov} or, equivalently, by the interaction with the d-electrons. The increasing binding strength towards the left in the series can simply be related to a decrease in the occupancy of the anti-bonding hydrogen - metal d levels as the d-band occupancy decreases. To second order in the interaction V_{ad}, the interaction is simply [11,15] $-2(1-f)V^2_{ad}/(C_d-\varepsilon_a)$, where C_d is the center of the d-bands, ε_a the renormalized hydrogen atomic level, and f the degree of filling of the d-band. The $(1-f)$ behaviour is very pronounced in fig.2. Similar trends are expected for other electro-negative adsorbates producing a 'closed shell' well below the Fermi level when embedded in a homogeneous electron gas.

3.2. Bond lengths

Whereas ΔE_{cov} determines the trends in binding energies along the transition metal series, it varies very weakly with distance outside a surface, as shown

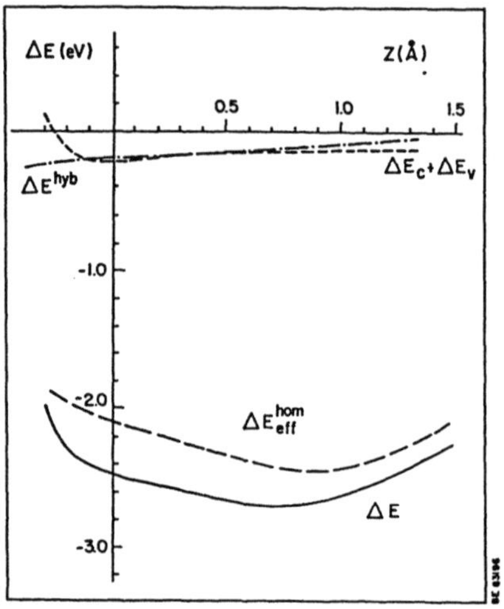

Fig.3. Binding energies of a hydrogen atom as a function of distance outside the fourfold coordinated center site on a Ni(100) surface. Both the total binding energy ΔE_{tot}, and the various components in (2) are shown. ΔE_{cov} is split up into first ($\Delta E_C + \Delta E_V$, see also ref.[18]) and higher order (ΔE^{hyb}) terms. From ref.[15].

in fig. 3. The bond length is thus determined mainly by the requirement that the surface electron density is equal to the optimum value. This, for instance, means that the bond length increases rapidly with coordination number, as illustrated in table 1, where bond lengths and vibrational frequencies for hydrogen and oxygen on Ni(100) and (111) are shown.

3.3. Vibrational excitations

It is clear from fig. 3 that the perpendicular vibrational frequency $\omega_\perp$ is also determined by $\Delta E^{hom}_{eff}(\bar{n}_0)$. It is then easy to show that in the harmonic approximation [15]

$$\omega_\perp \approx \frac{1}{\sqrt{M}} \sqrt{\frac{d^2 \Delta E^{hom}_{eff}(\bar{n}_0)}{d\bar{n}_0^2}} \; \frac{d\bar{n}_0}{dz} \quad , \tag{4}$$

where M is the atomic (reduced) mass and z denotes the direction perpendicular to the surface. For a given atom, $\omega_\perp$ is thus mainly determined by the perpendicular derivative of the surface electron density. $\omega_\perp$ is therefore

Table 1
Comparison of theoretical and experimental bond lengths and perpendicular frequencies (harmonic approx.) for H and O chemisorbed on Ni(100) and (111)

	Coordination number	Bond length (Å)		Vibrational frequency (meV)	
	Theory	Theory	Exp.	Theory	Exp.
H/Ni(100)	4	1.91		76	74
H/Ni(111)	3	1.81	1.84	131	139
O/Ni(100)	4	1.83	1.98	30	(36)
O/Ni(111)	3	1.76	1.88	62	72

For O/Ni(100), the experimental frequency shown is the 'static lattice' value [17,30]. For experimental references, see refs.[15,17]

given basically by the bond angle and thereby by the coordination number and the substrate geometry, as shown in table 1.

Parallel to the surface, $\Delta E^{hom}_{eff}(\bar{n}_0)$ only gives rise to a corrugation of the chemisorption well. Differences between different sites, parallel vibrational frequencies, and surface diffusion energies are therefore given by ΔE_{cov} [18]. Generally, this gives rise to rather small activation barriers, as shown in fig. 4 for hydrogen on Ni(100). The hydrogen mobility is thus expected to be large and tunneling is an important migration mechanism at low temperatures, as is also observed experimentally in one case [19]. The energy bands calculated by solving the Schrödinger equation for a proton in the potential of fig. 4 is shown in fig. 5 [20]. The bandwidth of excited states is seen to be substantial. This may be of smaller importance at high coverages. The very anharmonic nature of the potential parallel to the surface is expected to persist at higher coverages, though. The second excited band is basically the perpendicular vibration [20] and the energy position of this band is reasonably well given in the harmonic approximation used above.

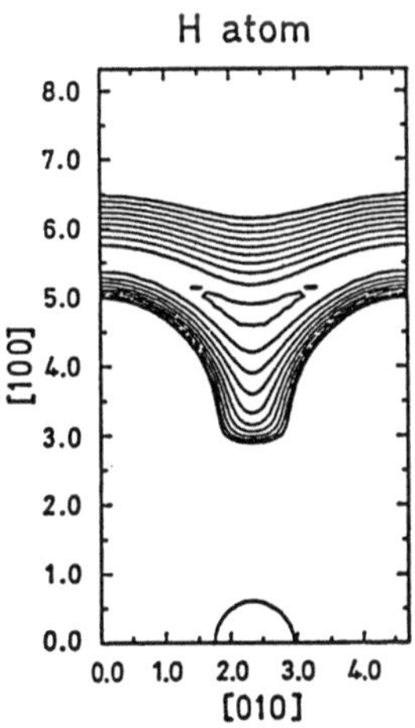

Fig.4. Contours of constant potential energy for hydrogen outside a Ni(100) surface. The contours are shown in a cut perpendicular to the surface through the two- and fourfold coordinated sites. The contours are shown in steps of 0.05 eV, the lowest value being -2.7 eV. From ref.[20]

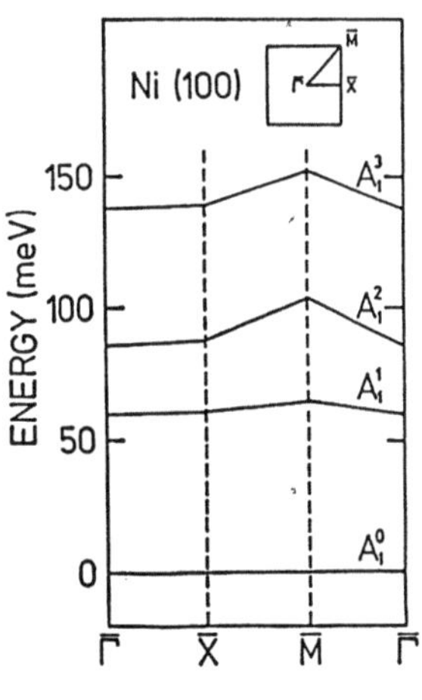

Fig.5. The band structure for hydrogen chemisorbed on the Ni(100) surface shown along the high-symmetry directions indicated in the insert. Only the states belonging to the A_1 representation of the C_{4v} point group are shown. The zero of energy is the ground-state energy (-2.6 eV) at the Γ point. This includes a zero-point energy of 0.1 eV. In the inset the Brillouin zone has been rotated 45° relative to the convention used in fig.4. From ref.[20]

3.4. Migration into the metal

The migration into the metal is again dominated by the density-dependent term $\Delta E_{eff}^{hom}(\bar{n}_0)$ or, equivalently, by the kinetic energy repulsion. In fig.6 an example of a potential energy variation for an oxygen atom moving into a Ni(111) surface is shown. The huge activation barrier is a consequence of the large electron density in the first metal layer. If the metal atoms are allowed to relax away from the oxygen atom, the electron density decreases and the barrier decreases. The interstitial electron density inside a transition metal is larger than the optimum value in fig.1. The energy of the oxygen (or hydrogen) atom inside the metal is therefore basically linearly dependent on the interstitial electron density [11]. As seen in fig.6, the energy can be reduced considerably by relaxing the nearest-neighbour metal atoms outwards from the oxygen (hydrogen) atom. The cost of relaxing a surface atom is smaller than the cost for a bulk atom and a subsurface site is therefore typically more stable than one further into the bulk [17].

It should be mentioned that the outwards relaxations of the metal lattice induced by the oxygen atoms inside the metal gives rise to a very strong attractive oxygen-oxygen interaction [17]. Oxygen atoms close to each other can share the cost of relaxing the nearest-neighbour metal atoms. This is consistent with the strong tendency for island formation observed in oxidation reactions [21].

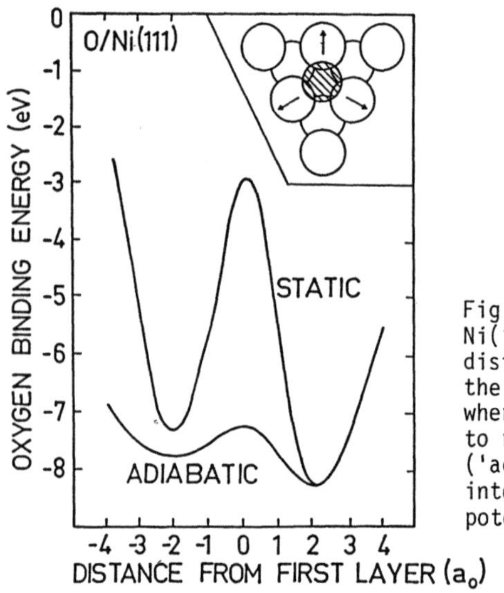

Fig.6. Energy of oxygen outside a Ni(111) surface as a function of distance from the first layer. Both the static lattice potential and one where the Ni atoms have been allowed to relax to give the minimum energy ('adiabatic') are shown. The Ni-Ni interactions are described by a pair potential. From ref.[17]

4. MOLECULAR ADSORPTION, CATALYSIS

For molecular adsorption and reactions on surfaces, even the evaluation (2) can be rather demanding. In studying the *trends* in molecular binding energies and activation energies, it can therefore be more instructive to apply (1) directly to get energy differences. The starting point is the molecule outside one surface and (1) then gives the energy-change when the surface is modified somehow [22,23].

4.1. *Poisoning and promotion*

As an example consider the interaction between adsorbing molecules and pre-adsorbed atoms. Generally, electro-negative pre-adsorbates like Cl, S or O are known to decrease the adsorption rate of molecules like CO, N_2 or H_2 [24], whereas electro-positive pre-adorbates like Li, Na or K increase it [25]. Experimentally, this effect has been related to a decreased/increased interaction of the molecule with the surface due to the presence of the co-adsorbate [24,25]. Exceptions to this picture are NH_3 and H_2O adsorption where electro-negative co-adsorbates *increase* [26] the adsorption strength and the alkali's *decrease* it [27].

In terms of (1), a co-adsorbed atom can change the interaction energy of the adsorbing molecule with the surface in two ways. The co-adsorbate induced electrostatic potential $\delta\phi_0$ will interact with the molecule. This is described by the first term. The second describes both the direct interaction due to an overlap between the molecule and co-adsorbate orbitals as well as the indirect one mediated through the surface electrons. In the following, only the first one will be considered. This does not, of course, imply that the second one is unimportant.

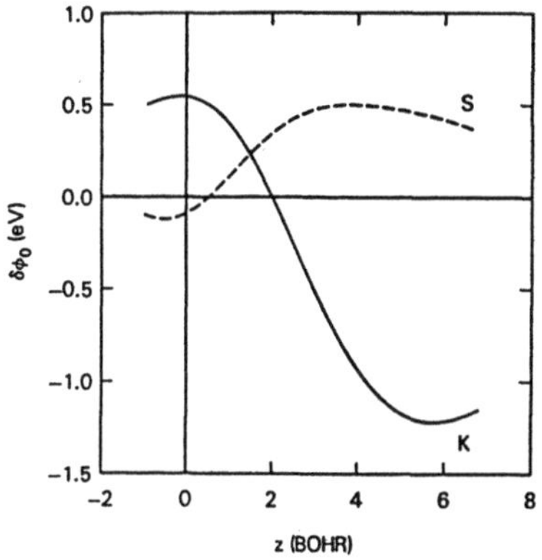

Fig.7. Variation of the self-consistent electrostatic potential $\delta\phi_0$ perpendicular to a jellium surface for S and K. The variation is shown 5 bohr from the adsorbed atom in the direction parallel to the surface. This corresponds to a typical adsorbate-adsorbate separation on a metal surface.

In fig.7 the self-consistently screened electrostatic potential $\delta\phi_0$ due to an electro-positive and an electro-negative atom chemisorbed on a jellium surface is shown. It is clear that a molecule like CO, which extracts electrons from the surface (into the anti-bonding $2\pi^*$ orbital) will be stabilized by nearby electro-positive co-adsorbates and de-stabilized by electro-negative co-adsorbates. Molecules like NH_3 or H_2O which, when adsorbed, have a large internal charge transfer towards the surface,will react in an opposite way. The electrostatic term alone can thus explain the trends observed experimentally [22,28].

The size of the effect has been estimated [28]. For CO/Na adsorption on Fe(110), and increase in the CO-metal interaction of 0.06 - 0.3 eV is found [22]. This lies well within the experimentally determined values [25].

5. Conclusions

The effective medium theory, which is a method for comparing total interaction energies of an atom or molecule with different solids has been described. For atomic chemisorption, detailed potential energy surfaces can be calculated very easily starting from the atom in a homogeneous electron gas. Furthermore, the properties of the interaction potential can be related to the properties of the surface and atom in question. The approach can also be used to compare interaction or activation energies in more complicated situations. Again, the simplicity of the approach allows an initial understanding of some of the parameters determining the reactivity and catalytic activity of surfaces.

Acknowledgements

Without the collaboration with Bulbul Chakraborty, S. Holloway, N. Lang, B. Lundqvist, M. Manninen, R. Nieminen, P. Nordlander and M. Puska, this work would not have appeared.

References

1. J.R. Schrieffer, J. Vac. Sci. Technol. 13, 335 (1976)
2. J.K. Nørskov, J. Vac. Sci. Technol. 18, 420 (1981)
3. For a recent review, see e.g. Theory of the Inhomogeneous Electron Gas, ed. S. Lundqvist and N.H. March (Plenum Press, New York, 1983)
4. T.H. Upton and W.A. Goddard, Phys. Rev. Lett. 42, 472 (1979)
5. P.J. Feibelman and D.R. Hamann, Sol. St. Commun. 34, 215 (1980)
6. C. Umrigar and J.W. Wilkins, to be published.
7. N.D. Lang and A.R. Williams, Phys. Rev. B18, 616 (1978)
8. B.I. Lundqvist, O. Gunnarsson, H. Hjelmberg and J.K. Nørskov, Surf. Sci. 89, 196 (1979)
9. J.K. Nørskov, A. Houmøller, P. Johansson and B.I. Lundqvist, Phys. Rev. Lett. 46, 257 (1981)
10. J.K. Nørskov and N.D. Lang, Phys. Rev. B21, 2136 (1980); M.J. Stott and E. Zaremba, Phys. Rev. B22, 1564 (1980)
11. J.K. Nørskov, Phys. Rev. B26, 2875 (1982)
12. O.K. Andersen, H.L. Skriver, H. Nohl and B. Johansson, Pure Appl. Chem. 52, 93 (1979); O.K. Andersen, in: The electronic structure of complex systems, NATO Advanced Study Institute, ed. W. Temmerman and P. Phariseau (Plenum Press, New York, 1982)
13. The most comprehensive calculations are those of M.J. Puska, R.M. Nieminen and M. Manninen, Phys. Rev. B24, 3037 (1980)
14. C. Umrigar, M. Manninen and J.K. Nørskov, to be published
15. P. Nordlander, S. Holloway and J.K. Nørskov, Surf. Sci. 136, 59 (1984)
16. D.M. Newns, Phys. Rev. 178, 1123 (1969)
17. B. Chakraborty, S. Holloway and J.K. Nørskov, to be published.
18. Close to the metal atoms there is a third contribution to the interaction energy, not included in (2), describing the interaction with the metal cores [15]. This term can in some cases contribute to energy differences along the surface, in particular on top of a surface atom. In most other cases it is negligible [15,17]
19. R. DiFoggio and R. Gomer, Phys. Rev. B25, 3490 (1982)
20. M.J. Puska, R.M. Nieminen, M. Manninen, B. Chakraborty, S. Holloway and J.K. Nørskov, Phys. Rev. Lett. 51, 1081 (1983)
21. See, for example, P.H. Holloway, J. Vac. Sci. Technol. 18, 653 (1981)
22. J.K. Nørskov, S. Holloway and N.D. Lang, Surf. Sci. 137, 65 (1984)
23. S. Holloway, B.I. Lundqvist and J.K. Nørskov, Proc. of the 8th Int. Congr. on Catalysis (Berlin, BRD, 1984)
24. See, for example, D.W. Goodman, R.D. Kelley, T.E. Madey and J.T. Yates, J. Catal. 63, 226 (1980); E.I. Ko and R.J. Madix, Surf. Sci. 109, 221 (1980)

25. See, for example, G. Ertl, S.B. Lee and M. Weiss, Surf. Sci. 114, 527 (1982); G. Brodén, G. Gafner and H.P. Bonzel, Surf. Sci. 84, 295 (1979)
26. D. Lachey, M. Surman and D.A. King, Vacuum 33, 867 (1983)
27. M. Grunze, F. Bozo, G. Ertl and M. Weiss, Appl. Surf. Sci. 1, 241 (1978)
28. N.D. Lang, S. Holloway and J.K. Nørskov, to be published
29. P.J. Feibelman and D.R. Hamann, Phys. Rev. Lett. 52, 61 (1984); R.W. Joyner, J.B. Pendry, D.K. Saldin and S.R. Tennison, Surf. Sci. 138, 84 (1984)
30. S. Andersson, P.-A. Karlsson and M. Persson, Phys. Rev. Lett. 51, 2378 (1983), and private communication

Applications of Force Analysis to Interactions Between Oxygen Atoms and Surface Atoms on the Al(111) and Mg(0001) Surfaces

Chikatoshi Satoko

Institute for Molecular Science, Okazaki, Aichi 444, Japan

1. Introduction

The local density functional method has been successfully applied to studies of electronic structures of molecules, solids and surfaces. Geometries such as the position of the adsorbates on the surface and the relaxation of the surface may be given by the calculation of the total energy. However, the geometrical optimization based on the total energy only is a laborious task for systems consisting of more than a few atoms. On the other hand, a direct force calculation method is useful in estimating reaction paths and equilibrium positions of the systems with many degrees of freedom. The stable arrangement can be automatically obtained by the movement of every atom in the direction of the forces acting on the atoms.

Recently we developed the direct force method [1] which is based on the linear combination of the atomic orbitals X-alpha method. The force acting on an atom is defined by the negative of the gradient of the total energy with respect to the atomic coordinate.

After some manipulations, we can get the formula of the total force which is decomposed into the following pair interactions between the atoms. The first term is the electrostatic force acting on the μ-th atom due to the τ-th atom:

$$\vec{f}^{\,\tau}_{\mu 1} = Z_\mu \vec{E}_\tau(\vec{R}_\mu) - \int \rho_\mu(\vec{r}) \vec{E}_\tau(\vec{r}) dv \,. \tag{1}$$

The electrostatic field $\vec{E}_\tau(\vec{r})$ is induced by the electronic charge and the nuclear charge of the τ-th atom. This force is long-ranged for the ionic materials. The electronic charge of the μ-th atom in the molecule or solid is defined by the Mulliken scheme:

$$\rho_\mu(\vec{r}) = \sum_i^{\mathrm{occ}} \psi_{i\mu} \psi_i \,, \tag{2}$$

where ψ_i is the i-th molecular orbital given by the linear combination of the atomic orbitals and $\psi_{i\mu}$ projected on the μ-th atom. By definition of the Mulliken scheme the overlap charge between the atoms is equally distributed on both of them. This force is derived from the assumption that the Mulliken charge distribution does not change for the infinitesimal displacement of the atom coordinate. The second term is due to the change of the Mulliken charge distribution such as the charge transfer, the charge polarization and the charge relaxation:

$$\vec{f}^{\,\tau}_{\mu 2}=\sum_i^{occ}\int\left[\frac{\partial\psi_{i\mu}}{\partial\vec{r}}(h-\epsilon_i)\psi_{i\tau}-\frac{\partial\psi_{i\tau}}{\partial\vec{r}}(h-\epsilon_i)\psi_{i\mu}\right]dv\ , \qquad (3)$$

where the term h is the one-electron Hartree-Fock-Slater hamiltonian and the term ϵ_i the orbital energy of the i-th molecular orbital. We call this force the deformed-charge force. The final term is due to the X-alpha exchange-correlation potential-field:

$$\vec{f}^{\,\tau}_{\mu 3}=\tfrac{1}{4}\int\left[\frac{d\rho_\mu}{d\vec{r}}\rho_\tau-\frac{d\rho_\tau}{d\vec{r}}\rho_\mu\right]V_{xc}(\vec{r})\rho^{-1}(\vec{r})dv\ . \qquad (4)$$

These forces can be calculated by using the self-consistent solution of the one-electron HFS-equation. The force acting on the atom is given by the summation of these three terms.

The decomposed forces for the AlO and Si_2 molecules are shown in figure 1(a) and (b), respectively, as a function of the bond distance. The force f acting on the oxygen in figure 1(a) is attractive at a long bond distance and become repulsive at the distance less than about 3 a.u., which agrees fairly well with the experiment [2]. This force is decomposed into the electrostatic force f1, the deformed-charge force f2, and the exchange force f3. At the equilibrium distance the force f1 is attractive, while the force f2 is repulsive. On the other hand the force f1 for the Si_2 molecule in figure 1(b) is repulsive at the equilibrium position 4.4 a.u. and the deformed-charge force f2 is attractive. The ionic material AlO is bonded by the electrostatic force f1 and the covalent one Si_2 by the deformed-charge force f2.

We have already applied these force analyses to problems such as oxygen chemisorption on the Al and Mg surfaces [3],

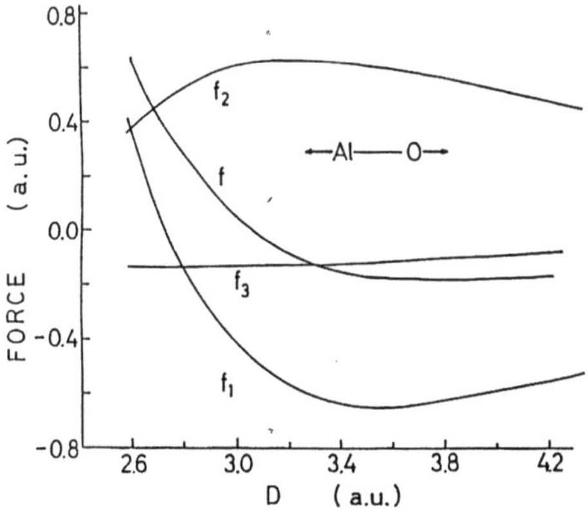

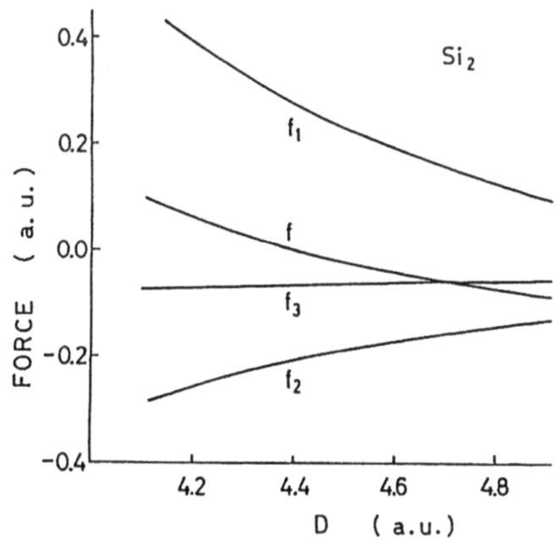

Fig.1 Calculated pair forces f and its components f_1, f_2 and f_3 for the molecules (a) AlO abd (b) Si_2

hydrogen dissociation on the metal surface[4],metal atom chemisorption on the silicon[5],and the silicon 7×7 structure[6]. Here we discuss the oxidation mechanism of the Al and Mg metals.

2. Experimental Facts

Oxidation process on the metal surfaces have been studied by many experimental methods such as LEED,XPS,AES,ELS and EXAFS[7-11]. BACHRACH and the others suggested from the experiments of EXAFS[7] that oxygen molecules are chemisorbed as molecules or dissociated atoms depending upon the oxygen pressure and the substrate temperature. At lower pressure than 2×10^{-7} Torr and at low temperature the oxygen molecule is undissociatively chemisorbed on the Al(111) surface. At higher pressure and at low temperature the oxygen molecule is dissociated into oxygen atoms which are chemisorbed at the centered threefold hollow site, while during the adsorption of the first 1/3 monolayer at room temperature,dissociated oxygen atoms penetrate through the Al(111) surface. When the coverage reaches 1 monolayer, the oxygen atoms form 1×1 overlayer at the three-fold hollow positions. They estimate the height of the oxygen as 2.0 a.u. and 1.3 a.u.,repectively,for the chemisorbed molecule and the dissociated atom. The position of the chemisorbed oxygen molecule is higher than that of the chemisorbed oxygen atom. As the coverage increases,some oxygen atoms are absorbed below the Al surface and an oxidation process begins.

On the other hand, NAMBA and the others[12] suggested that oxygen molecules on Mg(0001) surface are dissociated and absorbed below the surface plane. When the coverage increases, the assembly of the oxygen on the surface grows up in the shape of islands. The ordered overlayer structure was not observed.

To understand these chemisorption processes and the microscopic origin of the interactions between the adatom-surface and the adatom-adatom,we calculate the force acting on the oxygen and metal atoms as a function of the oxygen positions.It sheds light on the difference as to whether the dissociated oxygen atoms on the Al and Mg surfaces are absorbed or not,and whether they are assembled in island or not.

3. Interaction between the Oxygen Atom and the Metal Atoms

Firstly,we study the interaction between the dissociated oxygen atom and the surface atoms. A dissociated atom on the Al(111) and Mg(0001) surface is adsorbed at the centered threefold hollow site on the surface. To study the process ,we choose the model cluster consisting of an oxygen atom and four metal atoms,three from the first and one from the second layer,as shown in figure 2. The size of this cluster is small,but we can get the correct properties such as the chemical bonds.

The calculated pair forces between the oxygen atom,the first layer Al(1) and the second layer Al(2) are shown by the solids curves in figure 3 as a function of the height of the oxygen.

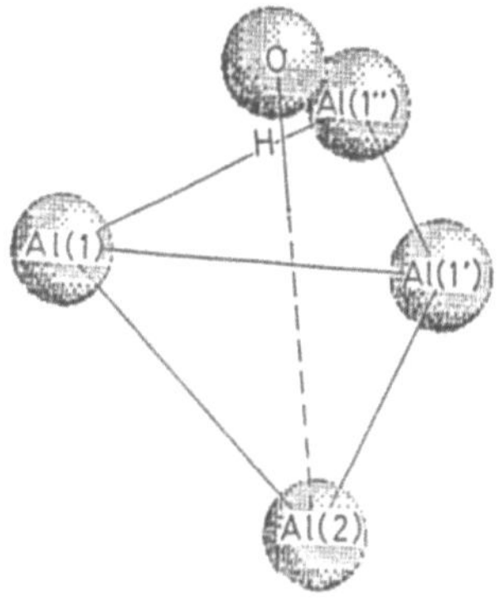

Fig.2 Geometry of the model cluster O/Al(1)$_3$Al(2) for the (111) surface

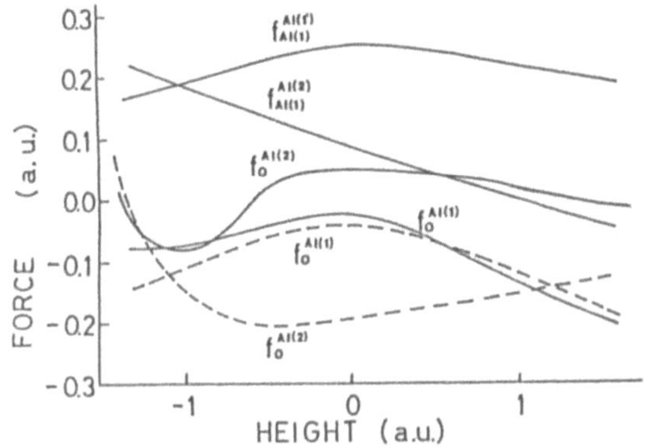

Fig.3 Effective pair forces among the O,the Al(1),and the Al(2) (after [1])

The dashed curves show the pair forces of the free AlO molecule corresponding to the bond distance between the O and the Al. The positive force is defined to be repulsive in the direction of the pair bond. The minus sign of the height means the oxygen is placed under the surface.

The pair force between the oxygen and surface Al(1) is always negative i.e. attractive at any height of the oxygen. The dependence of this force on the height is quantitatively similar to that of the interaction between the Al and the O in the free AlO molecule as shown in figure 3.

The pair force between the oxygen and the bulk Al(2) is ,however,different from the pair force in the AlO molecule. The pair force between the O and the Al(2) is repulsive when the oxygen is located near the surface,and changes to be attractive when it is more than 0.6 a.u. below the surface. It becomes repulsive again when the oxygen approaches the Al(2) and comparable to that in the AlO molecule.

The pair force between the first layer atoms Al(1) and Al(1') does not change,as this interaction is not under the influence of the bonding with the oxygen atom. On the other hand the pair force between the first layer Al(1) and the second layer Al(2) atoms changes to be more repulsive with the invasion of the oxygen atom into the surface.

These forces are calculated on the Al-Al bond distance fixed at the bulk values. When the oxygen atom approaches the surface,the surface Al atoms are distorted into the outward direction because of the repulsion between the Al(1) and Al(2) atoms as shown by figure 3.

We also calculate the total energy as a function of the height of the oxygen and find two stable positions of the chemisorbed oxygen atom. One is above the surface(H=1.0 a.u.) and the other below the surface(H=-1.1 a.u.). This result agrees with the experiment of the photoemission spectra [10]. With an increase in temperature the oxygen atom gains the energy necessary to penetrate the surface so that the photoemission peak from the Al 2p core state shifts to the positions of the bulk-like aluminum oxide. The activation energy for this change has been obtained from analysis of the peak intensity as about 1 eV. This value is comparable to the present calculated barrier height 0.84 eV.

On the other hand in the case of the oxygen chemisorption on the Mg(0001) surface the interaction of the oxygen with the first layer Mg(1) and the second layer Mg(2) in the cluster is found to be similar to the interaction of the oxygen with the Mg in the MgO molecule. The pair interactions of the first layer Mg(1) with the other first layer Mg(1') and with the second layer Mg(2) don't become so repulsive with the invasion of the oxygen as those of the Al(1) in figure 4 do. The force acting on the oxygen atom above the surface is always towards the surface. The oxygen atom is absorbed below the surface without activation energy. The position at which the force acting on the oxygen atom is vanishing has been found at the depth of 1.3 a.u.. The force acting on the first layer Mg(1) is always very small in respect to the position of the oxygen atom. The force acting on the second layer Mg(2) changes from pushing down to pulling up towards the surface at the oxygen height of -1.0a.u.

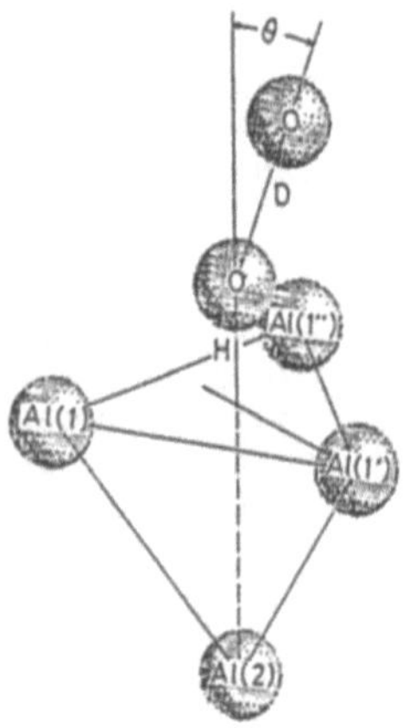

Fig.4 Geometry of the model cluster $O_2/Al(1)_3Al(2)$

The behavior of these pair forces can be understood in the following way. The orbital of the chemisorbed oxygen atom on the Al surface interacts with the 3p orbitals of the Al atoms. The three 3p orbitals of the Al(1) atom are bonded with those of the surrounding Al atoms. The direction of these 3p lobes changes with the approach of the oxygen to the surface because of their antibonding with the O 2p orbital. The approach of the oxygen tends to break the bonding between the surface atom Al(1) and the inside atom Al(2) and make the new bonding between the oxygen and surface Al(1) atoms.

The bonding between the Mg atoms is mainly formed of the 3s electrons. Therefore,the bonding property of the Mg atoms is not directional as compared with that of the Al atoms. The bonding of the Mg surface atoms is more insensitive to the approach of the oxygen to the surface as compared with the case of the Al surface. The oxygen on the Mg surface forms the bond with the Mg surface atoms without the bond break between the Mg atoms.

4. Interaction between the Oxygen Atoms on the Surface

In this section we discuss the interaction between the oxygen atoms on the surface. The model cluster as shown in figure 4 is chosen to study whether the chemisorption of an oxygen molecule is possible or not. We calculate the forces acting on the lower oxygen(Ol) and the upper oxygen(Ou) as a function of the height H, the bond distance between the Ol and Ou, and the azimuth angle θ . Figure 5 shows the force acting on the upper oxygen atom as a function of the height of the lower oxygen atom. The oxygen molecule is assumed to be upright, namely, θ=0.

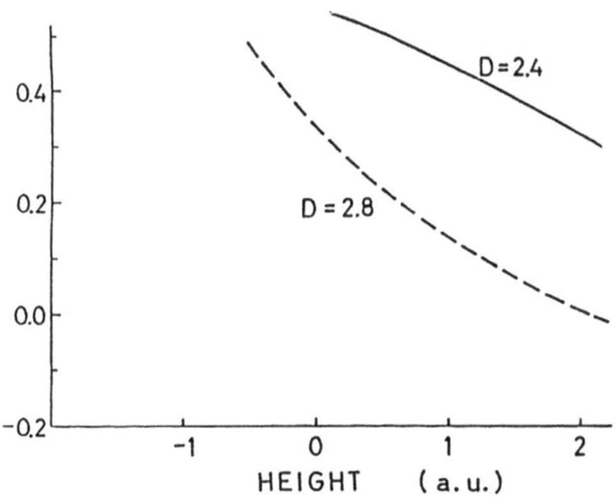

Fig.5 Force acting on the upper O of the figure 4 as a function of the height of the lower O

The solid curve in figure 5 shows the force with the distance of the oxygen atoms fixed at 2.4 a.u. which is the calculated bond distance of the free oxygen molecule. The force is so repulsive that the molecule is not chemisorbed on the surface. When the distance between the oxygen atoms is lengthened to the value 2.8 a.u., the force acting on the upper oxygen atom becomes zero at the position of the height 2.0 a.u. of the lower oxygen, as shown by the dashed curve in figure 5. The force acting on the lower oxygen atom is also found to be zero at that position. The adsorbed position of the molecule is further than that of the single oxygen atom.

When the lower oxygen atom penetrates the surface and is located around the position -1.0 a.u., the force acting on the upper oxygen on the surface is ,as shown in figure 5, so repulsive that the upper oxygen atom is not chemisorbed at the upper site above the absorbed lower oxygen. The repulsive force is due to the decrease of the electrostatic force between the lower oxygen and upper oxygen atoms.

In the case of the Mg surface we obtain from the calculated forces that the lower oxygen is absorbed and the upper oxygen atom is chemisorbed on the hollow site above the absorbed lower oxygen atom(H=-1.0 a.u. and D=2.7 a.u.).

We calculate the force acting on the upper oxygen in the declined molecule and find that the geometry of the molecule on the Al surface is declined by the angle of about 5 degrees on the surface. This arrangement is easy to interact with the Al atom. The bonding π orbital of the oxygen molecule becomes nearer to the first layer Al atom. However,the activation barrier(less than 0.1 eV) for the dissociation is found to be so small that the adsorbed molecule is dissociated into the atoms by the collision of the other atoms or the thermal energy.

The dissociated oxygen atoms are chemisorbed on the next nearest neighbor sites. When the coverage reaches 1 monolayer ,the arrangement forms 1×1 structure. So we study whether the dissociated atoms are chemisorbed on the nearest neighbor sites or not. We choose the cluster consisting of four first-layer metals,three second-layer metals and two adsorbed oxygen atoms as shown in figure 6(a) and (b) and calculate the forces acting on the oxygen atoms as a function of the oxygen heights. The arrows of figure 6(a) and (b) show the forces acting on the oxygen atoms at the heights H_1=1.0 a.u. and H_2=1.0 a.u..

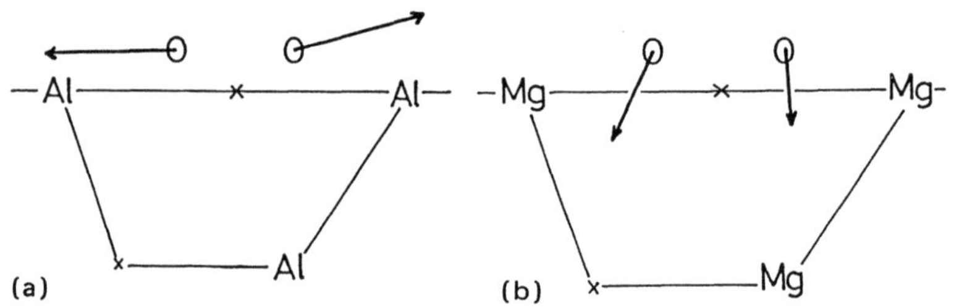

Fig.6 The forces (denoted by arrows) acting on the oxygen atoms of the heights 1.0 a.u. on (a) the Al and (b) the Mg surfaces. The geometries show the side view of the model cluster $O_2/M(1)_4M(2)_3$

The interaction between the oxygen atoms at the nearest neighbor positions is repulsive for the Al surface. While on the Mg surface the oxygen atoms penetrate into the surface from the nearest neighbor sites. Even in the case of the position where one is in the inside and another on the surface we find that the oxygen atoms for the Al surface repels each other and those on the Mg surface move into the direction of the invasion.

Whether the interaction between the nearest neighbor adatoms is repulsive or attractive is due to the following reasons. When the first adatom makes the surface metal-metal bonds weak, the second adatom is not chemisorbed at the nearest neighbor site of the first adatom because of the bond saturation of the metal atoms. In the case of the chemisorption of the oxygen atom on the Al(111) surface, the interaction between the Al atoms is weakened because of the bonding between the oxygen and Al atoms as shown in figure 3. On the other hand, the interaction between the Mg atoms is not so repulsive as to break the bond with the chemisorption of the oxygen atom. The interaction of the former case is repulsive, while the latter is not repulsive.

5. Conclusion

In summary, we lay stress again on the importance of the bond-switching mechanism in the chemisorption processes. Both bond making and bond break always occurs in the chemisorption and absorption processes. Generally, there are three possibilities according to the order of their forces.

Firstly, when the force for the bond-making is weak, then the atom or molecule is weakly adsorbed on the surface. This case occurs in the oxygen molecule chemisorption on Al surface. Secondly, when the force for the bond-making exceeds the force for the bond-break, then the atom is adsorbed on the surface. This is a frequent occurrence in the chemisorption process. In this case another atom is not adsorbed at the nearest neighbor site of the adsorbed atom, because metal orbitals to make bonds with the another atom are already used for the bond with the adsorbed atom. The case of the oxygen atom on the Al surface is this one. The observed phenomena [13] that nearly all geometries of the atom chemisorption on the metal(001) surface are not primitive 1×1 but centered 2×2 and primitive 2×2 structures are also due to the same reason. Thirdly, when the force for the bond-break is weak, then the chemisorbed atom is absorbed under the surface. The chemisorption process on the Mg surface is the third case. In this case the other atom of a molecule is chemisorbed or absorbed at the nearest neighbor sites.

Numerical calculations were carried out at the Computer Center of the Institute for Molecular Science on a HITAC M-200 computer.

References

1. C.Satoko, Phys.Rev.B30,1754(1984).
2. K.P.Haber and G.Herzberg, Constants of diatomic molecules(Van Nostrand, Princeton,1979).
3. C.Satoko, Chem.Phys.Letters 83,111(1981).
4. C.Satoko and M.Tsukada, Surf.Sci.134,1(1983).
5. S.Ohnishi and C.Satoko, to be published.
6. M.Tsukada and C.Satoko, to be published.
7. R.Z.Bachrach, G.V.Hansson and R.S.Bauer, Surf.Sci.109,L560(1981).
8. V.Mavitinez, F.Soria, M.C.Munoz and J.L.Sacedon, Surf.Sci.128,424(1983).
9. I.P.Batra and L.Kleinman, J.Electron Spectrosc. Relat.Phenom.33,175(1984).
10. S.A.Flodstrom, C.W.B.Martinson, R.Z.Bachrach, S.B.M.Hagstrom and R.S.Bauer, Phys.rev.Letters 40,907(1978).

11. R.L.Strong,B.Firey,F.W.de Wette and J.L.Erskine, J.Electron Spectrosc.Relat.Phenom.,29,187(1983).
12. H.Namba,J.Darville and J.M.Gilles, Surf.Sci.108,446(1981).
13. M.A.Van Hove,THE NATURE OF THE SURFACE CHEMICAL BOND ed.by T.N.Rhodin and G.Ertl(North-Holland, Amsterdam,1979) p.305.

Part II

Optical Processes

Surface Effects on Linear and Nonlinear Dynamics of Polaritons

Eiichi Hanamura

Department of Applied Physics, University of Tokyo
Hongo, Bunkyo-ku, Tokyo 113, Japan

Two effects of surfaces on linear and nonlinear dynamics of polaritons are discussed. First, the additional boundary condition is studied for the population dynamics of the polaritons in the film medium. Second, the optical bistability due to the polariton-excitonic molecule system is discussed by paying attention to the surface effects in the film.

1. Introduction

Two important effects of surfaces on polariton dynamics are discussed.

First, there are several cases in which the particle description is more suitable, so that the additional boundary conditions are developed in the particle picture for normal incidence of laser light with frequency larger than the longitudinal exciton in the film medium [1]. Then it is shown that while upper-branch polaritons are by a few orders of magnitude more densely excited than lower ones with large wave number in the semi-infinite medium, they are almost equalized in the film medium, because the latter are accumulated in the medium although the formers escape easily from both the surfaces. Both the polaritons are reflected into lower or upper ones with opposite wave numbers in sign so that the correlation effect among them is induced. These are discussed refering to the observations by Kuwata and Nagasawa [2].

Second, when the film medium is irradiated by intense and coherent laser light, the transmitted light shows the hysteresis against change of the incident laser power. This is called an optical bistability and originates from combined effects of the nonlinear dynamics of polaritons and the feedback at the surfaces [3]. Here it is shown that the microscopic nonlinear polarization is enhanced to the macroscopic scale by the surface effect. The nonlinear effect of the polaritons comes from two origins: the coherent (virtual) polarization of the excitonic molecule (EM) and the real (incoherent) excitation of the EM. When the virtual process is dominantly used for the optical bistability, we can switch between the on- (the high transmitting) and the off- (the low one) states in an order of pico-second [4]. On the other hand, it takes an order of subnano-second in the optical bistability due to the incoherent real excitation of the EM.

2. Additional Boundary Conditions

When an incident plane wave with a frequency ω_2 larger than the longitudinal exciton ω_L impinges normally on a semi-

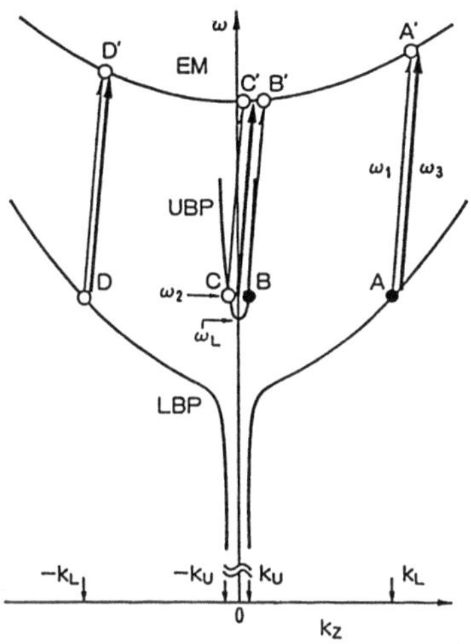

Fig.1. Dispersion curves of lower- and upper-branch polaritons (LBP and UBP) and excitonic molecule (EM). A- and B-polaritons are excited by excitation laser (ω_2) and reflected into C- and D-polaritons at the rear surface. One of these four polaritons is depopulated selectively by the pump laser (ω_3) and detected by the probe light (ω_1)

infinite dispersive medium, it will excite two polariton modes in the medium, i.e., the lower-branch polariton A and the upper one B in Fig.1. Let the vacuum be located at z>0, the medium at z>0. Then in the left half-space, there will be incident and reflected waves E_{in} and E_{re}. In the medium, there are two right-running waves E_A and E_B. The Maxwell boundary conditions give two relations: $E_{in}+E_{re}=E_A+E_B$ and $E_{in}-E_{re}=r_AE_A+r_BE_B$, where $r_A\equiv k_A/k_0$ and $r_B\equiv k_B/k_0$ are refractive indices of k_A- and k_B- polaritons with $k_0\equiv\omega_2/c$. An additional boundary condition (ABC) is required to determine E_A, E_B and reflectivity $R=|E_{re}/E_{in}|^2$ as three unknown values should be determined. This ABC originates from the boundary condition and the reflection effect of the exciton at the surface. The exciton wave function for the center-of-mass motion takes the form:

$$\psi_{\mathbf{q}}(\mathbf{r}) = [e^{iq_zz} + R(\mathbf{q})e^{-iq_zz}]Q(z)\exp[i(q_xx+q_yy)]. \tag{1}$$

We may take the simple form $R(\mathbf{q})=-1$ and $Q(z)=1$ ($z\geq 0$) and 0 ($z<0$) from a physical point of view. In terms of this wave function, the susceptibility $\chi(\mathbf{r},\mathbf{r}')$ is evaluated [5]. Then the polarization $P(\mathbf{r})$ and the internal field $E(\mathbf{r})$ are determined from the following equations:

$$P(\mathbf{r}) = \int \chi(\mathbf{r},\mathbf{r}')E(\mathbf{r}')d\mathbf{r}', \tag{2}$$

$$-(\nabla^2+k^2)E(\mathbf{r}) = 4\pi k_0^2P(\mathbf{r}), \tag{3}$$

where $k_0\equiv\omega_2/c$. Here we Fourier-decompose the internal field in the form: $E(\mathbf{r})=\sum_i E_i\exp(ik_iz)$ for $z>0$, and insert it into eqs.(2) and (3). Then the polarization is evaluated from eq.(2). Comparing the same Fourier-components in eq.(3), we have Ewalt extinction theorem:

$$r_i^2 \equiv (k_i/k_0)^2 = 1 + 4\pi\chi(k_i, k_0), \tag{4}$$

and the ABC:

$$(r_A^2 - r_0^2)E_A + (r_B^2 - r_0^2)E_B = 0, \tag{5}$$

where $r_0^2 = 1+4\pi\chi_0$, χ_0 is the background susceptibility and $\chi(k_i, k_0)$ is the Fourier-transformed susceptibility. The former means that the internal fields are finite only when they satisfy the dispersion relation of the polariton, and the latter shows that the polarization with the wave vector of the naked exciton vanishes [5]. The ABC eq.(5) is rewritten in the particle picture by assuming the equality of energy flows in both the pictures [1]:

$$\frac{(r_i E_i)^2}{4\pi} v_{pi} = \hbar\omega_i n_i^0 v_{gi}, \quad i=A,\ B. \tag{6}$$

Here v_{pi} and v_{gi} are the phase velocity and the group velocity of the polariton with the wave vector k_i (i=A,B). They are given, respectively, as the ratio ω_i/k_i and the derivative $d\omega_i/dk_i$ and are quite different from each other. For example, at $\hbar\omega$=3.3209eV in CuCl, v_{pA}=c/40 and v_{pB}=c/0.92 [6] while $v_{gA}=10^{-4}c$ and $v_{gB}=5\times10^{-4}c$ [7]. Then the concentration ratio $n_A^0/n_B^0=5.6\times10^{-3}$ for the semi-infinite medium.

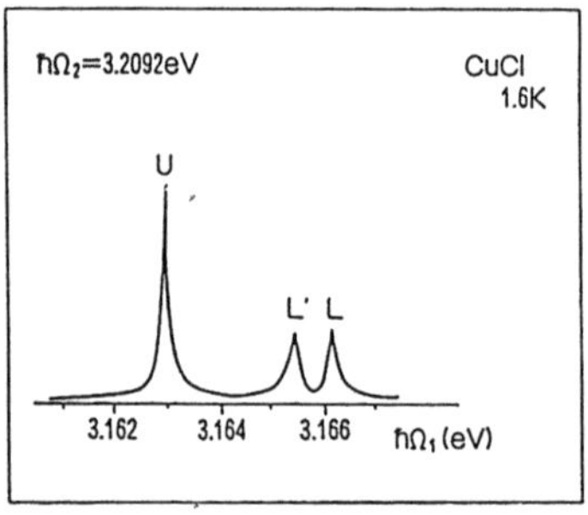

Fig.2. Signal proportional to the polariton concentrations at U (+U'), L and L' as a function of the probe laser frequency ω_1 (ref.[2])

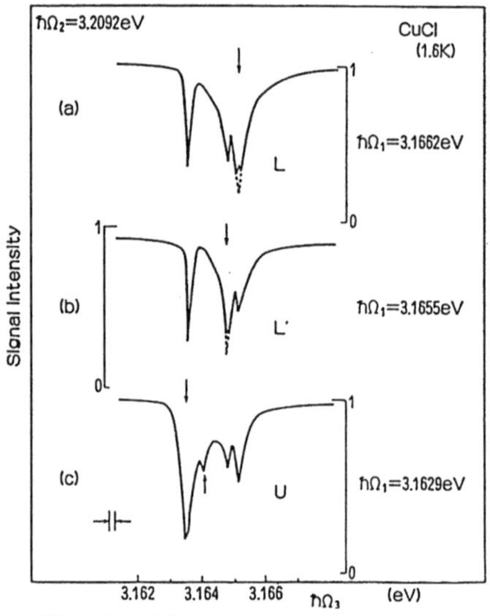

Fig.3. Signal proportional to the degree of the polariton concentration change $(n_i - n_i^0)/n_i^0$ due to the pump laser (ω_3) irradiation: (a) i=A (L-polariton), (b) i=D (L'-polariton) and (c) i=B+C (U-polariton) (ref.[2])

Two interesting facts were observed for the film medium [2]. Polariton densities at A and B are observed as induced absorption from these polaritons into the EMs A' and B' in Fig.1. The same information is obtained as the polarization rotation of the linearly polarized probe light ω_1 when the polaritons A and B are excited by one circularly polarized light ω_2. First, it was found that the lower-branch polaritons with the forward (A) and backward (D) propagating wave vectors denoted by L and L' are populated with the same order of magnitudes as the upper-polariton U as shown in Fig.2. Here the transitions of two upper-polaritons from B to B' and from C to C' are almost degenerate in energy and are not resolved in these observations. Second, the A-polariton concentration was observed to be reduced when the pump laser ω_3 is resonant, not only to the transition A→A' but also to those B(C)→B'(C') and D→D' as Fig.3a shows. The similar behavior was observed for the L'- and U-polariton concentration as shown by Figs.3b and 3c. Now these two puzzles are resolved by solving the polariton dynamics in the film [1].

3. Polariton Dynamics in Film Medium

We assume that the exciton wave function has nodes at both the surfaces z=0 and ℓ. Then the wave number q_z has discrete values $q_z=\pi n/\ell$ (n:integer). In the present case, the laser spectrum width is much larger than the discrete energy separation of the polaritons in the 20μm film medium, so that we may treat the polariton wave vector as the continuum variables. Under such a situation, the polariton dynamics will be described more appropriately in terms of the polariton density rather than the polarization itself. When an A-polariton packet hits the rear surface, it dissolves into C- and D-polaritons and the transmitting field E_{tr}. We rewrite the Maxwell's conditions: $E_A+E_C+E_D=E_{tr}$ and $r_A(E_A-E_D)-r_BE_C=E_{tr}$ and the ABC: $(r_A^2-r_0^2)(E_A+E_D)+(r_B^2-r_0^2)E_C=0$ in the particle picture eq.(6). Here the relations $\mathbf{k}_D=-\mathbf{k}_A$ and $\mathbf{k}_C=-\mathbf{k}_B$ were used. Then we have the reflection ratio from these relations:

$$n_D^0 = u_{DA}^2 n_A^0, \tag{7a}$$

$$n_C^0 = u_{CA}^2 \left(\frac{v_{pA}}{v_{pB}}\right)\left(\frac{v_{gA}}{v_{gB}}\right) n_A^0, \tag{7b}$$

$$n_{tr}^0 = u_{EA}^2 \left(\frac{v_{gA}}{v_{pA}}\right) n_A^0, \tag{7c}$$

where

$$u_{DA} = -\frac{(r_A+r_B)(r_Ar_B+r_A-r_B-r_0^2)}{(r_A-r_B)(r_Ar_B+r_A+r_B+r_0^2)}, \tag{8a}$$

$$u_{CA} = \frac{2r_A(r_A^2-r_0^2)}{(r_A-r_B)(r_Ar_B+r_A+r_B+r_0^2)}, \tag{8b}$$

$$u_{EA} = \frac{2r_A(r_A+r_B)}{r_{AB}+r_A+r_B+r_0^2}. \tag{8c}$$

We repeat similar calculations for the cases when the B-polariton packet hits the rear surface and when the C- and D-polariton packets hit the front surfaces. Then the polariton dynamics are derived by taking into account the four effects: first the excitation I_A and I_B of the A and B polaritons at the front surface, second the reflection and transmission U_{ij} of these polaritons at both the surfaces, third the relaxation Γ_i and finally the pumping G_i of the specified polaritons into the EMs:

$$\dot{n}_A = I_A+U_{AC}n_C+U_{AD}n_D-\Gamma_A n_A-G_A n_3 n_A, \tag{9a}$$

$$\dot{n}_B = I_B+U_{BD}n_D+U_{BC}n_C-\Gamma_B n_B-G_B n_3 n_B, \tag{9b}$$

$$\dot{n}_C = U_{CA}n_A+U_{CB}n_B-\Gamma_C n_C-G_C n_3 n_C, \tag{9c}$$

$$\dot{n}_D = U_{DA}n_A+U_{DB}n_B-\Gamma_D n_D-G_D n_3 n_D. \tag{9d}$$

Here n_3 is the polariton density at ω_3 excited by the strong pump laser, and $G_i=8\gamma_{i'}\omega_3\mu'^2/\hbar[\omega_{i'}+(\omega_i+\omega_3-\omega_{i'})^2]$ with $\hbar\omega_{i'}$ the EM energy at $k_{i'}=k_i+k_3$, $\gamma_{i'}$ the transverse relaxation constant and μ' the transition dipole moment between the exciton and the EM. U_{ij} is expressed in terms of u_{ij}, phase and group velocities of polaritons i and j. For example, $U_{CA}=u_{CA}^2(v_{pA}v_{gA}/v_{pB}v_{gB})(2v_{gA}/\ell)$. $\Gamma_A=\Gamma_A^0+2v_{gA}/\ell$ with the intrinsic relaxation constant Γ_A^0. The first experiment observes polariton densities n_i under no pumping $G_i=0$. The second one gives the population change $(n_i-n_i^0)/n_i^0$ under the ω_3 pumping of the specified polariton into the EMs [2]. First, for the incident power $100kW/cm^2$ and the film thickness $20\mu m$, we have the polariton densities: $n_A^0=n_D^0=0.6\times10^{16}cm^{-3}$, $n_B^0=1.7\times10^{16}cm^{-3}$ and $n_C^0=1.0\times10^{16}cm^{-3}$, where the polariton decay rate was assumed $\Gamma_A^0=10^7sec^{-1}$. These polaritons are almost equally populated, because the densely excited upper-branch polaritons escape easily from the surfaces while the lower ones are accumulated in the medium. Second, we evaluate from eqs.(9) the degrees of the A-polariton decrease

(a) under $A\rightarrow A'$ pumping,

$$\frac{n_A-n_A}{n_A^0} = -\frac{G_A n_3/\Gamma_A}{1+G_A n_3/\Gamma_3} = -0.95, \tag{10a}$$

(b) under $D\rightarrow D'$ pumping,

$$\frac{n_A-n_A^0}{n_A^0} = -\frac{U_{AD}n_D^0}{\Gamma_A n_A^0}\,\frac{G_D n_3/\Gamma_D}{1+G_D n_3/\Gamma_D} = -0.95, \tag{10b}$$

(c) under B+C $\rightarrow$B'+C' pumping,

$$\frac{n_A-n_A^0}{n_A^0} = -\frac{U_{AD}U_{DB}n_B^0+\Gamma_D U_{AC}n_c^0}{U_{AD}U_{DB}n_B^0+\Gamma_D U_{AC}n_c^0+\Gamma_D I_A}\,\frac{G_C n_3/\Gamma_C}{1+G_C n_3/\Gamma_C} = -0.58. \tag{10c}$$

These degrees of the population decrease are comparable to the observations of Fig.(3a). Therefore we may conclude that these two puzzles have been solved by the present calculation [1]. The used material constants of CuCl are listed: r_3=2.54 at $\hbar\omega_3$=3.165eV and $G_i n_3=2(\mu' E_3/\hbar)^2/\pi\gamma_{i'}$ with $\hbar\gamma_{i'}$=0.2meV and μ'=10 . Here the transition dipolemoment of the exciton μ is obtained from the L-T splitting 5.7meV$\equiv\Delta_{LT}=4\mu^2/a_0^3$ with the exciton Bohr radius a_0=6.7A.

4. Nonlinear Polariton Dynamics

In this section, we discuss the surface effect on the nonlinear dynamics of polaritons. When the film medium is irradiated by coherent and intense laser light, the transmitted light shows the hysteresis against change of the incident laser power. This is called an optical bistability and originates from combined effects of the nonlinear dynamics of polaritons and the feedback at the surfaces. The film medium composes the Fabry-Perot resonator. The transmitted power I_t is expressed in terms of the phase-difference δ for one round trip and the finesse F as

$$I_t = \frac{1}{1+F\sin^2(\delta/2)}\,I_i. \tag{11}$$

Here F is determined by the reflectivity R and the attenuation δ' as $F=\{4R/(1-R)^2\}\exp(-\delta')$ and δ is composed of two terms proportional to linear refractive index n_0 and the nonlinear part n_2E^2 as

$$\delta = 4\pi n_0\ell/\lambda+(4\pi\ell/\lambda)n_2E^2. \tag{12}$$

The transmittance oscillates as a function of δ. The transmitted power I_t is related to the internal field E as $I_t=(1-R)E^2$. Therefore I_t is obtained as solutions of the coupled equations of (11) and (12) as a function of I_i. I_t shows a hysteresis as shown in Fig.4. It is noted that the refractive index change n_2E^2 of an order of 10^{-3} is enough to induce the phase-change of an order of π because of the enhancement factor $4\ell/\lambda\sim10^3$ for ℓ=5μm. Let us consider irradiating the crystal with a strong monochromatic laser light ω_0 at the nearly two-photon resonant excitation of an EM. Then the polariton with ω_0 and the corresponding wave vector $\mathbf{K}_0$, and the virtual excitation of an EM with $2\omega_0$ and $2\mathbf{K}_0$ are induced in the crystal effectively due to the effect of the

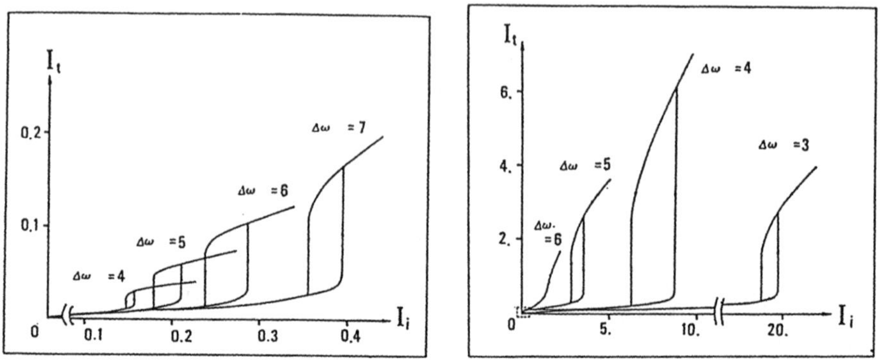

Fig.4. Optical bistability due to CuCl film. The normalized transmitted power I_t shows double hysteresis against the change of the normalized incident power I_i for various detuning Δω normalized by the transverse relaxation constant

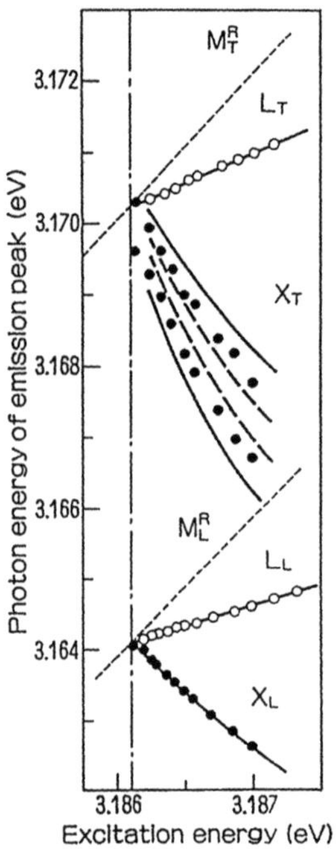

Fig.5. The band peak frequencies as a function of the incident frequency ω_0. M_T^R and M_L^R are the hyper-Raman peaks leaving behind the transverse and longitudinal excitons. Solid and dotted lines denote theoretical results and circles the observations

giant oscillator strength [8]. At the same time, the EMs are really formed on its dispersion $\omega_m(2\mathbf{K}_0)$. These three excitations compose the macroscopic condensed states. With these states as the initial state, we have several emission bands. In the lowest emission process, there were observed two hyper-Raman lines from the virtual EM and two luminescence bands from the really excited EMs, both leaving behind the transverse and longitudinal excitons [9]. Hyper-Raman lines denoted by M_T^R and M_L^R shift as the incident frequency ω_0 changes as shown in Fig.5 while luminescence lines are fixed. Hyper-Raman scattering and luminescence are diagrammatically represented by Fig.6a and 6b, respectively. As the next higher-order processes [10], two emission bands denoted by X-

Fig.6

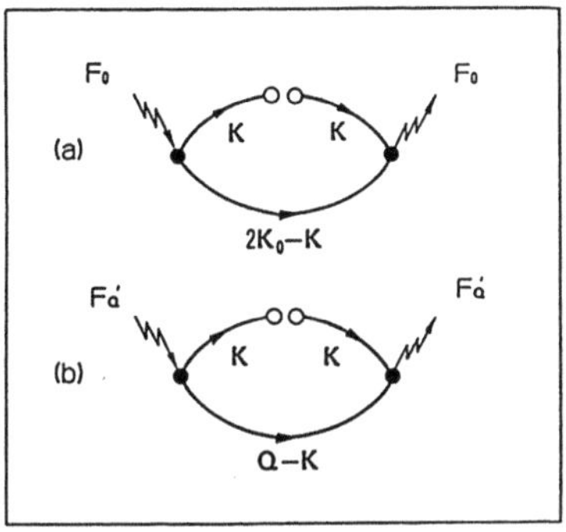

Fig.7

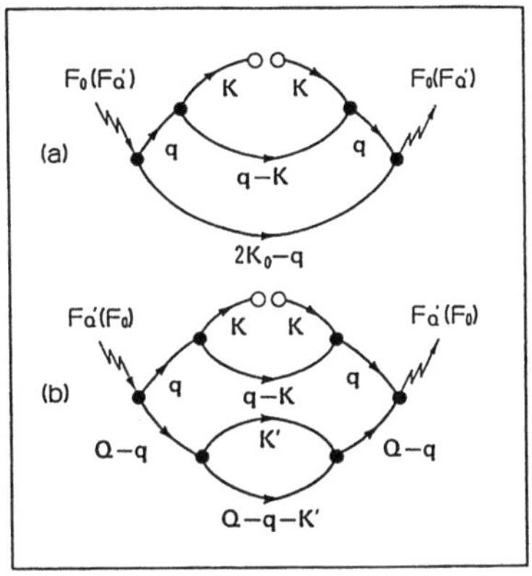

Fig.6. Diagrams describing the hyper-Raman scattering (a) and the luminescence (b) of the EM

Fig.7. (a) Diagram inducing the X_T- and X_L-bands, and (b) that inducing the L_T- and L_L-bands. It is noted that the polariton $\mathbf{q}$ is hybridized with the EM absorbing the coherent polariton $(\omega_0, \mathbf{K}_0)$

and L-bands were observed [11]. Two polaritons $\mathbf{K}$ and $2\mathbf{K}_0$-$\mathbf{K}$ are formed from a virtual EM of $2\omega_0$ and $2\mathbf{K}_0$, and then both polaritons interact with incident polaritons at ω_0 and $\mathbf{K}_0$ to form two EMs. One of these is resolved into two polaritons and one of these polaritons is observed as a X_T- and X_L-bands. This is represented by the diagram Fig.7a. L_T- and L_L-bands are observed when a real EM at $\omega_m(2\mathbf{K}_0)$ is annihilated according to the diagram of Fig.7b. The ω_0-dependence of these emission bands can explain the observed characteristics as shown in Fig.5. These higher-order emission processes are extremely enhanced and are made observable by two effects of the giant oscillator strength and the resonance [8]. Corresponding to the coherent (virtual) and incoherent (real) excitations of the EM, we have two mechanisms to induce the nonlinear polarization of the polariton inducing the optical bistability [12].

Now we pursue the dynamics of the expectation values of polariton operators α_F and β_B, forward and backward propagating with the wave vector $\mathbf{K}_0$ and $-\mathbf{K}_0$ in the film medium: $\alpha_F \equiv \langle a_{\mathbf{K}_0} \rangle \exp[i(\omega_0 t - \mathbf{K}_0 z)]$ and $\alpha_B \equiv \langle a_{-\mathbf{K}_0} \rangle \exp[i\{\omega_0 t - \mathbf{K}_0 (2\ell - z)\}]$. Under the coherent excitation, these expectation values are finite. Here we take into account the nonlinear coupling of these polaritons with polarizations associated with excitations of the polaritons into the EMs, the virtual polarization C of the EMs and the real creation N of the EMs. As there are $\mathbf{K}_0$ and $-\mathbf{K}_0$ polaritons in the film medium, we have several βs, Cs and Ns depending upon the combinations:

$$\beta_{F1} \equiv \langle b_{2\mathbf{K}_0} a^+_{\mathbf{K}_0} \rangle \exp[i(\omega_0 t - K_0 z)],$$

$$\beta_{F2} \equiv \langle b_0 a^+_{-\mathbf{K}_0} \rangle \exp[i(\omega_0 t - K_0 z)],$$

$$\beta_{B1} \equiv \langle b_{-2\mathbf{K}_0} a^+_{-\mathbf{K}_0} \rangle \exp[i\{\omega_0 t - K_0 (2\ell - z)\}],$$

$$\beta_{B2} \equiv \langle b_0 a^+_{\mathbf{K}_0} \rangle \exp[i\{\omega_0 t - K_0(2\ell - z)\}],$$

$$C_F \equiv \langle b_{2\mathbf{K}_0} \rangle \exp[2i(\omega_0 t - K_0 z)],$$

$$C_0 \equiv \langle b_0 \rangle \exp[2i(\omega_0 t - K_0 \ell)],$$

$$C_B \equiv \langle b_{-2\mathbf{K}_0} \rangle \exp[2i\{\omega_0 t - K_0(2\ell - z)\}],$$

$$N_F \equiv \langle b^+_{2K_0} b_{2K_0} \rangle, \quad N_0 \equiv \langle b^+_0 b_0 \rangle \quad \text{and} \quad N_B \equiv \langle b^+_{-2\mathbf{K}_0} b_{-2K_0} \rangle.$$

For example, α_F, β_{F1}, C_F and N_F obey the following equations (see ref.(12) for full details):

$$\dot{\alpha}_F = - v_g \frac{\partial}{\partial z} \alpha_F - ig(\beta_{F1} + \beta_{F2}), \tag{13a}$$

$$\dot{\beta}_{F1} = -(\gamma' + i\Delta_{mp})\beta_{F1} - ig|\alpha_F|^2\alpha_F + 2ig(|C_F|^2 + N_F)\alpha_F, \tag{13b}$$

$$\dot{C}_F = -(\gamma + i\Delta_m)C_F - ig\alpha_F^2, \tag{13c}$$

$$\dot{N}_F = - \gamma N_F + A_0|\alpha_F|^5/(|\Delta_m|^2 + \gamma^2 + \gamma|\alpha_F|^2). \tag{13d}$$

Here we take into account the effects of the transverse (γ ,γ') and longitudinal (γ) relaxation and the off-resonance $\Delta_{mp} \equiv \omega_m(2\mathbf{K}_0) - \omega_p(\mathbf{K}_0) - \omega_0$ and $\Delta_m \equiv \omega_m(2\mathbf{K}_0) - 2\omega_0$. The boundary conditions are assumed for the rear and front surfaces:

$$\sqrt{R}\alpha_F(z=\ell,t) = \alpha_B(z=\ell,t), \tag{14a}$$

$$\alpha_F(0,t) = \varepsilon_I(t) + \sqrt{R}\alpha_B(0,t)e^{i\delta}. \tag{14b}$$

Here $\varepsilon_I(t)$ means the supply from the incident laser field and is the deviation from $2\pi m$ of the phase-change for one round trip of the polariton. First, the stationary solution of the polariton density $|\alpha_F|^2$ is obtained as the solution of eqs.(13) and the equations of other β, C and N. The transmitted power $I_t \equiv (1-R)|\alpha_F|^2$ shows two step hysteresis as a function of the incident power $I_i \equiv |\varepsilon_i|^2$ for a reasonable detuning of an order of 0.5meV and a sample thickness of an order of 5μm. Examples are shown in Fig.4 for various detuning $\Delta\omega$s. The optical bistability at the lower incident power (<1 MW/cm^2) comes from the coherent, virtual polarization of the EM, while that at the higher one (∿10 MW/cm^2) is due to the incoherent, real excitation of the EM.

The dynamical response of these optical bistabilities is interesting not only from the physical point of view but also from the engineering point of view. There is an idea to use this optical bistability as an information processing and memory device in the optical computer. To overcome the present electric computer, we must be able to switch between the on- and off-states, i.e., the high and low transmitting states in an order of pico-second. When the system is operating under the incoherent excitation of the EM, it takes the lifetime of the EM, i.e., subnano-second to switch off. On the other hand, under the coherent virtual excitation of the EM, it can be switched on and off in an order of pico-second for the CuCl film of 5μm thickness [12]. In this case, the switching time is almost determined by the time of flight of the polariton in the medium. Furthermore, this coherent optical bistability is possible at a few orders of magnitude-weaker indicent laser than the other one. The optical bistability of this exciton-EM system was proposed in 1981 [4,13]. In 1983, the optical bistability due to this system of CuCl was observed in Strasbourg [14] and Arizona [15]. However, this was done almost under two-photon resonance of the EM. Then the holding power was $20MW/cm^2$, i.e., by a few orders of magnitude higher than the coherent one, and the switching time was limited by the measurement precision of an order of subnano-second. Therefore we may conclude that this is due to the incoherent process. The ratio of the contributions from the coherent process over the incoherent one increases as the degree of the off-resonance from the EM increases. Under reasonably off-resonant two-photon excitation of the EM (0.5meV for CuCl), we can make a coherent process dominant over the incoherent one [12].

In both of the linear and nonlinear polariton dynamics, we could show the important roles of the surfaces in the film medium.

References

1 E. Hanamura: Solid State Commun. 51, 213 (1984)
2 M.Kuwata and N. Nagasawa: Solid State Commun. 45, 973 (1983)
3 H.M. Gibbs, S.L. McCall and T.N.C. Venkatesan: Optics News/Summer, 6 (1979)
4 E. Hanamura: Solid State Commun. 38, 939 (1981)
5 R. Zeyher, J.L. Birman and W. Brenig: Phys. Rev. B6, 4613 (1972)
6 T. Mita, K. Sôtome and M. Ueta: Solid State Commun. 33, 1135 (1980)
7 Y. Masumoto and S. Shionoya: J. Phys. Soc. Jpn 51, 181 (1982)
8 E. Hanamura: Solid State Commun. 12, 951 (1973)
9 T. Mita, K. Sôtome and M. Ueta: J. Phys. Soc. Jpn 50, 134 (1981)
10 T. Tokihiro and E. Hanamura: Solid State Commun. 51, (1984).
11 T. Itoh, T. Katohno, T. Kirihara and M. Ueta: J. Phys. Soc. Jpn 53, 854 (1984)
12 T. Tokihiro and E. Hanamura: to be submitted to Solid State Commun.
13 S. Koch and H. Haug: Phys. Rev. Lett. 46, 450 (1981)
14 B. Hönerlage, J.Y. Bigot and R. Levi: In Proceedings of the Optical Bistability Conference, Rochester, 1983.
15 N. Peyghambarian, H.M. Gibbs, M.C. Gibbs, M.C. Rushford and D.A. Weinberger: Phys. Rev. Lett. 51, 1962 (1983)

Interference of Polariton Waves in a Thin Film of CuCl: Additional Boundary Conditions and the Effect of Selective Pumping

Kikuo Cho

Department of Material Physics, Faculty of Engineering Science
Osaka University, Toyonaka, Osaka 560, Japan

1. Introduction

It is well known that, in a spatially dispersive medium which is described by a dielectric function with an explicit dependence on wave vector k, as well as frequency ω, $\varepsilon(k,\omega)$, an electromagnetic wave with a given frequency can propagate in several different modes, i.e., with various wavelengths. At the boundary of this medium, in touch with a non-spatially dispersive medium (vacuum for example), there arises a problem of so-called "Additional Boundary Conditions (ABC)", since the number of the waves on both sides is, for a given frequency, larger than that of the Maxwell boundary conditions (MBC: Tangential components of E and H should be continuous.). This ABC problem, first noted by PEKAR [1], has attracted much attention for a long time from both theoretical and experimental points of view [2]. A study of interference pattern in a thin film has been considered to provide very useful information about such media. It is, however, only very recently that fine experimental results have become available [3,4]. They provide a good chance to sensitively check the various ABC theories proposed so far. Especially in CuCl, further experiments with the use of pumping light have been carried out to obtain information as to how the upper and lower branch polaritons are produced and mutually correlated through ABC [4,5]. These measurements have given rise to a new aspect of the ABC problem, and the theoretical interpretation of these pumping experiments is not yet established in spite of several proposals [6-8].

The purpose of this note is two-fold: [i] to show that a microscopic ABC theory of D'Andrea - del Sole type in the limit of very steep evanescent wave, which is equivalent to the Pekar ABC, can reproduce the interference pattern for a very thin (1500 A thick) sample almost completely, and [ii] to show that the effect of selective pumping on the interference pattern for the thin film can be understood semi-quantitatively in terms of a pumping-induced change in polariton dispersion, and that the pumping effect on a thicker (20 μm thick) sample should contain some additional, unknown physics.

2. Interference Pattern without Pumping

D'ANDREA - DEL SOLE [9] proposed a microscopic ABC theory for a simple Wannier exciton system with a flat boundary which act as an infinite barrier for both electron and hole. The new element in their theory is the inclusion of evanescent waves (of the form $\sim \exp[-PZ]$), as well as propagating waves, for the translational motion of an exciton in the exciton wave function for the semi-infinite space. They showed that the evanescent waves lead to an effect very similar to that of the so-called "exciton dead layer", the concept of which was introduced intuitively on the basis of a crude modelling of image potential near the surface [10].

In applying the above theory to our slab geometry, we restricted the number of possible evanescent waves to just one, and regarded the steepness constant P as an adjustable parameter, which is a slight modification from the view-point of the original theory. Assuming that $Pd \gg 1$ where d is the thickness of the slab, we have carried out the same procedure as D'Andrea - del Sole in calculating the electric susceptibility $\chi(R,R';\omega)$ of the slab and in solving the resulting Maxwell equation of integro-differential type. Due to the lack of space, we give here the minimal necessary results. The details will be reported elsewhere [11]. The wave components relevant to the calculation of interference effect are shown in Fig.1. E_i, E_r and E_t are incident, reflected and transmitted amplitudes, respectively, and E_j (E_j') the amplitude of the right (left) running bulk polariton of the j-th branch (j=1,2). All the bulk modes are accompanied by the evanescent waves near the both surfaces. For example, near Z=0, we have the total electric field of the form

$$E(Z) = \sum_{j=1}^{2} [\{ \exp(ik_jZ) + A_- \exp(-PZ) \} E_j + \{ \exp(-ik_jZ) + A_+ \exp(-PZ)\} E_j' \exp(ik_jd)] , \quad (1)$$

where

$$A_\pm = 2P (q^2 + P^2) / [(k_j^2 - q^2) (P \pm ik_j)] \quad (2)$$

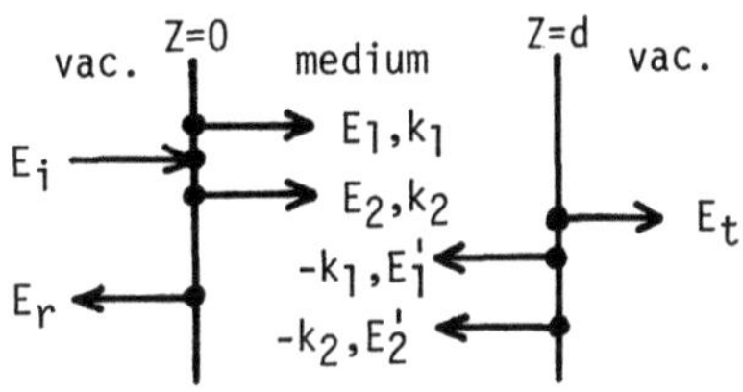

Fig.1 Relevant wave components. The phase of each amplitude E is defined at the point with a solid dot.

The wave vectors of the bulk modes $\{ k_j \}$ are the solutions of

$$\varepsilon(k,\omega) = (ck/\omega)^2 \quad (3)$$

for a given ω, and in order for the expression (1) to be the solution of the Maxwell integro-differential equation, the following condition

$$0 = \sum_{j=1}^{2} [B_- E_j + B_+ E_j' \exp(ik_jd)] / (k_j^2 - q^2) \quad (4)$$

must be fulfilled. This condition plays the role of ABC at Z=0, and

$$B_\pm = [\pm k_j + i (\delta - P)] / (P \pm i k_j) \quad . \quad (5)$$

The parameters q and δ are the known functions of ω, P and other material constants, which are defined in [9], eqs.(30,44). Since we have six boundary conditions (two sets of "two MBC and an ABC" on both surfaces), we can uniquely determine the reflectivity R and the transmittance T. Note that the effect of internal multiple reflection is automatically taken care of by this formalism up to the infinite order. For the numerical treatment, we used the following bulk dielectric function

$$\varepsilon(k,\omega) = \varepsilon_b + \varepsilon_b \Delta_{LT} / [\omega_X(k) - \omega - i\gamma] \quad (6)$$

where the exciton dispersion is characterized by the mass m as

$$\omega_X(k) = \omega_T + \hbar k^2/2m \quad , \tag{7}$$

and the numerical values are

$$\varepsilon_b = 5.59, \quad m = 2.23\, m_e, \quad \hbar\omega_T = 2.2023\ \text{eV}, \quad \Delta_{LT} = 5.7\ \text{meV} \quad . \tag{8}$$

The value of the film thickness d is easily determined by the number of the interference fringes between ω_T and ω_L ($= \omega_T + \Delta_{LT}$). Thus, the only adjustable parameters are P and γ. After several trials of fitting, we immediately notice that a constant value for γ never works. Since the ω-dependence of γ is a natural consequence of a theory where γ is regarded as the self-energy correction of the exciton due to all the possible scattering effects, we used a simplified model of the ω-dependence of the following form:

$$\hbar\gamma = \begin{cases} a & (\omega \le \omega_T) \\ a + b\,\hbar(\omega - \omega_T)/\Delta_{LT} & (\omega > \omega_T) \end{cases} \tag{9}$$

The best fit was obtained for the following values;

$$a = 0.02\ \text{meV}, \quad b = 0.04\ \text{meV} \tag{10}$$

As to the value of P, 1/P = 0.35 A, which is the value to be expected from the original theory [9] in the case of CuCl, works quite well, and the resulting spectra show little difference from those for 1/P = 0 [12]. Figure 2 shows the best fit results, to be compared with the measured spectra in [4]. It should be noted that not only the overall shape but also the rather delicate feature, such as the pairing behavior in transmittance spectrum and the alternating peak height in the reflectivity spectrum, both just below ω_L, could be reproduced by this theoretical treatment.

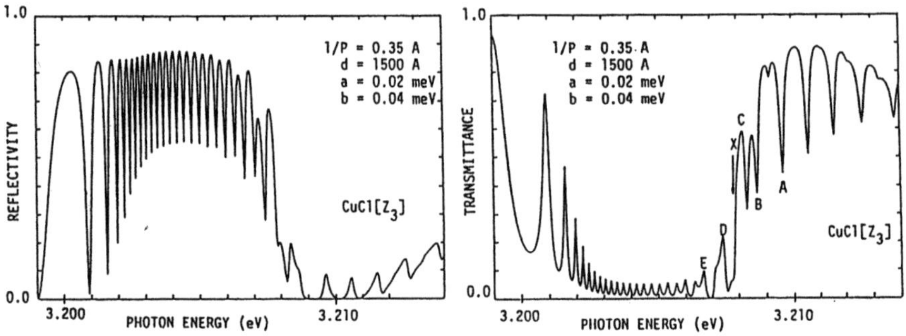

Fig.2 Calculated spectra of reflectivity and transmittance. The notations (A, B, C, X, D, E) in the latter correspond to those in Fig.3a,b.

Since 1/P = 0 corresponds to the Pekar ABC [9], our result shows the validity, in the case of CuCl, of the simple Pekar ABC, namely, "the resonant part of the polarization vanishes at surface, when summed over all the polariton contributions".

3. Effect of Selective Pumping

If one pumps the already existing polaritons up to the excitonic molecule (EM) state by the use of pumping light with appropriate frequency Ω, the relevant polariton state suffers from a non-linear change in its dispersion. Since the energy of EM depends also on its wave vector (the EM mass is nearly equal to 2m), each of the polariton states in Fig.1 has its own resonant pumping energy. The effect of pumping has been observed in CuCl as the modulation of R and T, or as the polarization rotation of the third (probe) light [4,5]. In this note, we analyze the pumping effect on R and T only. The effect on the polarization rotation will be considered elsewhere.

The effect of pumping could be described in the framework of dielectric formalism by considering an additional self-energy term that represents the effect of giant two photon absorption (GTA) to the EM states [13]. In the presence of pumping light of frequency Ω and wave vector Q, the dielectric function for the light (ω,k) can be written as

$$\varepsilon_p(k,\omega) = \varepsilon_b + \varepsilon_b\Delta_{LT} / [\omega_X(k) - \omega - i\gamma + \Sigma] \quad , \tag{11}$$

where, according to HAUG et al. [13],

$$\Sigma = \beta / [\omega + \Omega - \omega_M(k+Q) + i\gamma'] \quad . \tag{12}$$

The EM dispersion is assumed to be of the form

$$\omega_M(K) = \omega_M + \hbar K^2/2M \quad , \tag{13}$$

and β consists of the product of the matrix element of GTA and the population of the pumping light. The life time broadening of the EM state is expressed by γ'. Since the right hand side of eq.(11) can be rewritten as the sum of two simple poles as a function of ω, the system with pumping is equivalent to the two-oscillator (or two-"exciton") system. This leads to the existence of three polariton states for a given ω in the presence of pumping, and necessarily to a new ABC problem. Though it should be, in principle, treated again from a microscopic viewpoint, we skip the procedure and simply generalize the form of the ABC obtained above: Namely, since we already know that the Pekar ABC works best in this material in the absence of pumping, we assume an extended form of Pekar ABC in the presence of pumping. This means that the amplitude of each "exciton" state should vanish at each surface, when summed over all the polariton branches. The usefulness of this extended Pekar type ABC has already been tested in the analyses of reflectivity and light scattering intensity from multicomponent polariton systems [14].

The effective two-"exciton" system which leads to the dielectric function (11) is described by the following Hamiltonian matrix

$$H(k) = \hbar \begin{pmatrix} \omega_X(k) - i\gamma \;, & \sqrt{\beta} \\ \\ \sqrt{\beta} \;, & \omega_M(k+Q) - \Omega - i\gamma' \end{pmatrix} \tag{14}$$

and by the dipole matrix elements $(\sqrt{\varepsilon_b\Delta_{LT}}\ ,\ 0\ )$ for the two basis states of the Hamiltonian. In this case, the general argument of the extended Pekar ABC [15] leads to the following form of ABC's to be used:

$$0 = \sum_{j=1}^{3} [\, H(k_j) - \hbar\omega \,]^{-1}{}_{t1}\, E_j \quad , \quad (t=1,2) \tag{15}$$

where k_j's are the bulk polariton wave vectors in the presence of pumping.

Numerical result for the modulation spectra of transmittance T for the thin (d=1500 A) sample at several different ω is shown in Fig.3b together with the measured spectra, Fig.3a [4]. The following parameter values were used in addition to those in (8):

$$\hbar\omega_M = 6.3720 \text{ eV}, \quad M = 5.29\, m_e, \quad \hbar\gamma' = 0.20 \text{ meV}$$
$$\hbar\gamma = 0.05 \text{ meV (const.)}, \quad \hbar^2\beta = 1.0 \times 10^{-8} \text{ eV}^2 \quad . \tag{16}$$

The value for γ is taken constant in this calculation, since the region of ω in question is rather narrow. Fig.3a and Fig.3b show that this theory well reproduces the evolution of the pumping spectrum for T as one changes the frequency ω of the observation from the point A to E. The agreement turns out to be not only qualitative, but also semi-quantitative, if one compares the full scale units carefully. The only exception is at the frequency denoted as X. At this frequency which is almost equal to ω_L, not only the pumping spectrum but also the interference spectrum itself show rather poor agreement with the measurements. Several trials to overcome this difficulty, including those mentioned in [12], have been unsuccessful, and it remains open for future study.

The pumping effect on reflectivity at $\omega = \omega_L + 1.1$ meV/$\hbar$ for thicker samples is given in Fig.3c. The result for d = 20 μm has no resemblance (except for the order of magnitude of $|\Delta R|$) to the measured spectrum, Fig.5 in [5]. The use of smaller broadening parameters for thicker samples might be suggest-

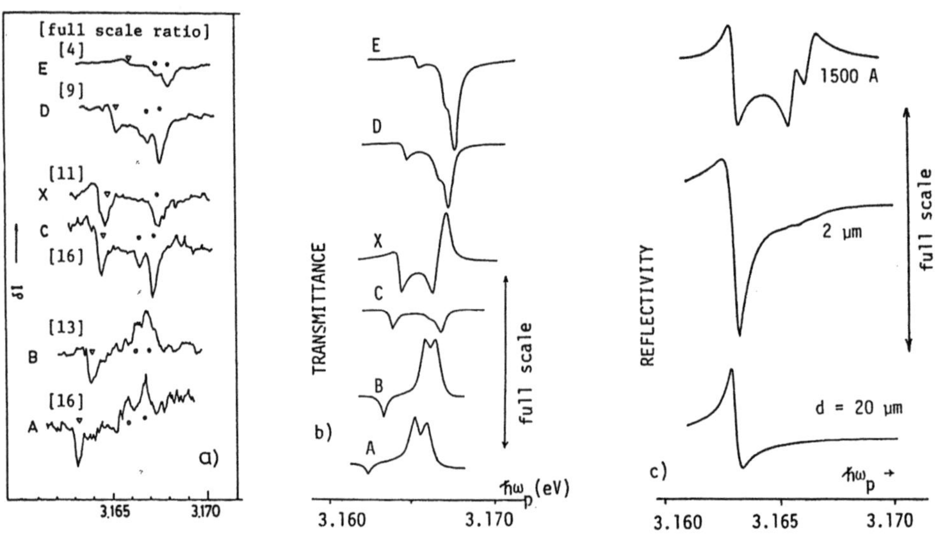

Fig.3 Pumping effect on transmittance for a thin sample (a, b) and on reflectivity (c). Different full scale units are used in the measured curves (a), from ref.[4]. The symbols A - E refer to the frequencies shown in Fig.2.

ed, since γ's for thinner samples may include more uncertainty of the polariton wave vector due to the size effect. This does not, however, bring about any essential improvement in the comparison between theory and experiment. Thus, we are led to the conclusion mentioned as [ii] at the end of introduction. All the theoretical proposals made so far to explain the unexpected pumping effect in the thicker sample [5] would encounter the difficulty: Any one of them, including the present one, cannot explain the effect of selective pumping for both thin and thick samples at the same time. Since the result for the thin sample can be understood semi-quantitatively by the present theory, the remaining problem is, in the author's opinion, to find out the unknown physics involved in the case of thick sample.

The author would like to thank Mr. M. Kawata, Osaka University, for his asistance with some of the calculations and Professor N. Nagasawa for providing the unpublished data about the full scale ratio used in Fig.3a.

References

1. S.I. Pekar: Sov. Phys. JETP 6, 785 (1958)
2. J.L. Birman: "Excitons" ed. E.I. Rashba and M.D. Sturge (North Holland, Amsterdam 1982) p.27
3. V.A. Kiselev, B.S. Razbirin, and I.N. Uraltsev: Phys. Stat. Solidi (b) 72, 161 (1975)
4. T. Mita and N. Nagasawa: Solid State Commun. 44, 1003 (1982)
5. M. Kuwata and N. Nagasawa: Solid State Commun. 45, 937 (1983)
6. M. Nakayama: Solid State Commun. 45, 821 (1983)
7. T. Takagahara: Solid State Commun. 47, 345 (1983)
8. E. Hanamura: Solid State Commun. 51, 697 (1984)
9. A. D'Andrea and R. Del Sole: Phys. Rev. B25, 3714 (1982)
10. J.J. Hopfield and D.G. Thomas: Phys. Rev. 132, 563 (1963)
11. K. Cho and M. Kawata: to be published
12. K. Cho: Fig.3(c) and the related text in ref.[4]
13. H. Haug, R. März, and S. Schmitt-Rink: Phys. Letters A77, 287 (1980), S. Schmitt-Rink and H. Haug: Phys. Stat. Solidi (b) 108, 377 (1981)
14. W. Dreybrodt, K. Cho, S. Suga, F. Willmann, and Y. Niji: Phys. Rev. B21, 4692 (1980),
    K. Cho, S. Suga, and W. Dreybrodt: "Physics in High Magnetic Fields" ed. S. Chikazumi and N. Miura, (Springer, Heidelberg 1980) p.190
    P. Fiorini, J.C. Merle, and M. Simon: Phys. Rev. B22, 4941 (1980)
    M. Yamane and K. Cho: Physics 117B & 118B, 377 (1983)
15. K. Cho: Solid State Commun. 27, 305 (1978)

Charge Transfer Instability in Optical Processes of Adsorbates

Kazumasa Shinjo and Satoru Sugano

The Institute for Solid State Physics, The University of Tokyo
7-22-1 Roppongi, Minato-ku, Tokyo 106, Japan

Dynamical calculations of the optical absorption, photoelectron emission and luminescence spectra of the system consisting of an adsorbate and a metal substrate are made with a particular reference to the effects of the critical line, across which the valence charge of the adsorbate is expected to change discontinuously to a static approximation. It is assumed that the system is described by a simple spinless Hamiltonian, and a time-dependent selfconsistent mean-field approximation is applied. The calculated absorption and photoelectron emission spectra show a broadening when the initial effective valence level of the adsorbate crosses the critical line. The broadening may be ascribed to a charge-transfer instability of the system. The results are favorably compared with those of recent experiments. It is pointed out that the relative positions of the neutral and the charge-transferred excited states of the adsorbate may be known by observing the luminescence from the adsorbate.

1 Introduction

Dynamical properties of the system consisting of an adsorbate and a metal substrate have been studied in connection with photoelectron[1,2] and optical-absorption[3] spectra by employing the spinless model Hamiltonian,

$$H = \varepsilon_c n_c + [\varepsilon_a - U_{ac}(1 - n_c)]n_a + \sum_k \varepsilon_k n_k + V\sum_k (C_a^\dagger C_k + \text{H.c.}) + (n_a + n_c - 1)\sum_p g_p(b_p^\dagger + b_p) + \sum_p \omega_p b_p^\dagger b_p, \tag{1}$$

where ε_c is the energy of the core level of the adsorbate, n_c the number operator of the core electron (n_c=0,1), U_{ac} the stabilization energy of the adsorbate valence electron by a core hole, ε_a the energy of the unoccupied adsorbate valence level which couples via V to the substrate described by eigenvalues ε_k, and n_a (n_k) the number operator of the adsorbate (substrate) valence electrons. Throughout this paper, the energy is measured from Fermi surface of the substrate, ε_F, which is taken to be zero. The operator $C_a^\dagger(C_k^\dagger)$ is the creation operator of the adsorbate (substrate) valence electrons. The surface plasmon with frequency ω_p, whose number operator is $b_p^\dagger b_p$, couples to the net charge of the adsorbate with the coupling strength g_p.

Recently static properties of a similar system have been studied[4] in detail by employing a more general Hamiltonian including spins and the Coulomb interaction between the adsorbate electrons. Within the adiabatic approximation, the presence of a critical line in a parameter space has newly been found, across which the valence charge of the adsorbate in the ground state changes discontinuously. Such a critical line for the Hamiltonian in eq.(1) with n_c=0 is shown by a solid bold line in Fig.1, where the adiabatic potential curve is given as a function of the plasmon coordinate and η =

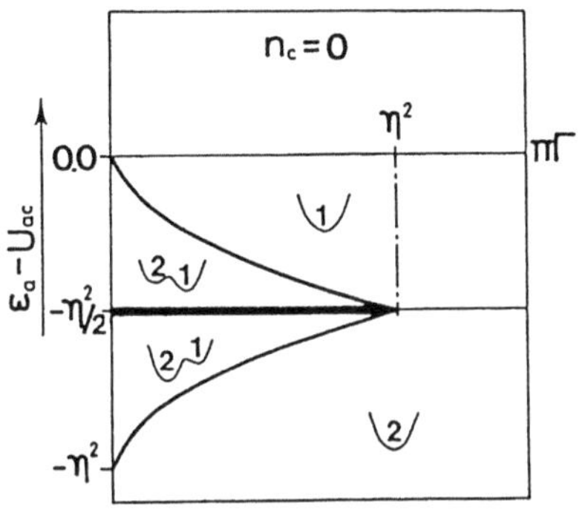

Fig.1 Adiabatic potential curves in the parameter space. The critical line is shown by the solid-bold line.

$(2\Sigma g_p^2/\omega_p)^{1/2}$. The minima numbered by 2 and 1, respectively, correspond to the neutral and positively ionized states of the adsorbate. Parameter Γ is defined as $\Gamma=\pi|V|^2\Sigma\delta(\varepsilon-\varepsilon_k)$, assumed to be constant.

The purpose of the present article is to summarize the results of dynamical calculations of the optical absorption, photoelectron emission and luminescence spectra of the system given by Hamiltonian (1) *with a particular reference to the effects of the critical line,* and to discuss their implications in connection with experiments.

2 Dynamical Treatment[5] and Charge Transfer Instability

In our treatments, we adopt the mean-field approximation[1],

$$n_a\sum_p g_p(b_p^\dagger+b_p) \to \langle n_a\rangle_t\sum_p g_p(b_p^\dagger+b_p)+n_a\sum_p g_p\langle(b_p^\dagger+b_p)\rangle_t - \langle n_a\rangle_t\sum_p g_p\langle(b_p^\dagger+b_p)\rangle_t, \tag{2}$$

where $\langle\ \rangle_t$ denotes the expectation value in the state at time t, for example, in state $|\psi(t)\rangle=\exp(-iH_f t)C_a^\dagger C_{ac}|i\rangle$ for the absorption and in state $|\psi(t)\rangle=\exp(-iH_f t)C_{ac}|i\rangle$ for the photoelectron emission. Here, H_f is the Hamiltonian for the final state obtained by putting $n_c=0$ in eq.(1), C_{ac} the annihilation operator of the core electron, and $|i\rangle$ the eigenstate for the Hamiltonian (1) with $n_c=1$.

For the absorption and the photoelectron emission, we assume that stabilization of the one-electron adsorbate valence level by the core-hole field occurs instantaneously at $t=0$ so that the effective energy level of the adsorbate valence electron at $t=0$ is given as $\tilde\varepsilon_a(t=0)=\varepsilon_a-U_{ac}$. Then, the core-field is getting screened by the excitation of the surface plasmon, and the effective energy level is pushed up as

$$\tilde\varepsilon_a(t)=\varepsilon_a-U_{ac}+\Sigma g_p\langle(b_p^\dagger+b_p)\rangle_t. \tag{3}$$

The position of this effective energy level relative to the Fermi level of the substrate determines the occupation number $\langle n_a\rangle_t\equiv n_a(t)$ of the adsorbate valence state. Since the screening potential $\Sigma g_p\langle(b_p^\dagger+b_p)\rangle_t$ is determined by $n_a(t)$, $\tilde\varepsilon(t)$ and $n_a(t)$ are determined selfconsistently at each time t in our treatments.

An example of the temporal change of $\tilde\varepsilon(t)$ and $n_a(t)$ in the case of the absorption is shown in Fig.2. In this example, the critical line has to be

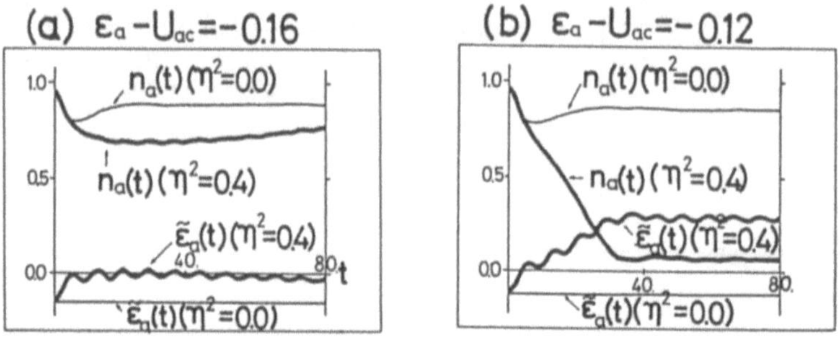

Fig.2 Calculated temporal changes of the effective energy of the adsorbate valence level, $\tilde{\varepsilon}_a(t)$, and its occupation number $n_a(t)$.

located at -0.2. However, since we assume the number of the substrate valence levels to be finite (40), the excitation of one electron into the valence states pushes up the Fermi level a little above zero. This relocates the critical line between -0.16 and -0.12 in the final state. Accordingly, Fig.2(a) and Fig.2(b) correspond to the cases where $\tilde{\varepsilon}_a(0)$ is below and above the critical line, respectively. In case $\tilde{\varepsilon}_a(0)$ is below the critical line, the screening potential $\Sigma g_p \langle (b_p^{\dagger}+b_p) \rangle_t$ cannot be large enough to push up $\tilde{\varepsilon}_a(t)$ above the Fermi level, consequently $n_a(t)$ stays near unity showing little charge transfer as seen in Fig.2(a). In case $\tilde{\varepsilon}_a(0)$ is above the critical line, $\tilde{\varepsilon}_a(t)$ can cross the Fermi level at some time t. Then, the adsorbate valence electron begins to flow into the substrate. This onset of charge transfer further increases the screening potential, accelerating the charge transfer until the adsorbate valence level becomes empty as seen in Fig.2(b). This is called a *charge transfer instability*.

3 Calculated Spectral Shapes

The numerical calculations are performed by assuming the presence of a single surface plasmon: its frequency denoted by ω_0 is set equal to 0.8, which is about 10 eV for Mo. η^2, twice the image force stabilization energy, amounts to 4 - 5 eV independent of substrate metals. Γ generally amounts to 1 - 2 eV for chemisorption and $\sim$0.1 eV for physisorption. All the energies of our system are scaled by ω_0.

The calculated spectral shapes of the absorption corresponding to the cases of $\tilde{\varepsilon}_a(0)$ below and above the critical line are shown in Fig.3(a) and Fig.3(b), respectively. The origin of the ω scale is determined so that the

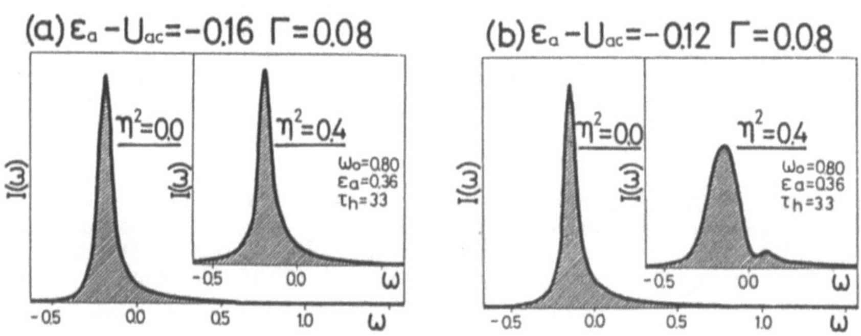

Fig.3 Calculated spectral shapes of the absorption.

zero of ω corresponds to the energy of the transition from the core level to Fermi surface. The spectral shapes shown in Fig.3(a) for $\eta^2=0$ and 0.4 and in Fig.3(b) for $\eta^2=0$ are nearly the same. They are sharp with a slightly broadened edge singularity[6,7] at the low-energy side and a small tail at the high-energy side. The broadening of the edge singularity is caused by the use of a finite discrete set of the energy levels and the finite lifetime τ_h of the core hole. The spectral shape in Fig.3(b) for $\eta^2=0.4$ is quite different. It is broad with an oscillatory long tail at the high-energy side. It shows *no edge singularity* at the low-energy side. The broadening is due to the charge transfer instability and may be interpreted as the lifetime broadening of the intra-atomic excitation.

The calculated spectral shapes of the photoelectron emission corresponding to the cases of $\tilde{\varepsilon}_a(0)$ above and below the critical line at -0.23 are shown, together with the temporal changes of $\tilde{\varepsilon}_a(t)$ and $n_a(t)$, in Fig.4(a) and Fig.4(b), respectively. The zero of ω corresponds to the one-electron transition energy from the core level when the plasmon coupling is absent. The $\tilde{\varepsilon}_a(t)$ and $n_a(t)$ curves show a charge transfer instability similar to that found in the absorption. The shapes in Fig.4 show a zero-plasmon line shifted by $\sim$+0.25 by the plasmon coupling and a plasmon satellite at the low-energy side of the zero line. The zero-line is broadened when $\tilde{\varepsilon}_a(0)$ crosses downwards the critical line as shown in Fig.4(b). The oscillatory tail at the low-energy side of the zero line in Fig.4(b) is quite similar to that at the high-energy side of the absorption line in Fig.3(b).

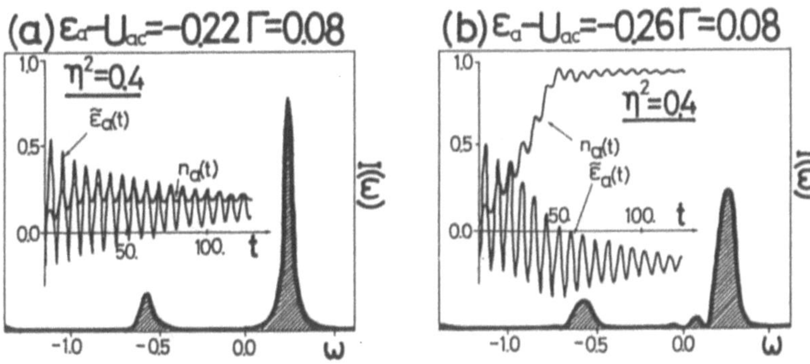

Fig.4 Calculated line shapes of the photoelectron emission, together with calculated temporal changes of $\tilde{\varepsilon}_a(t)$ and $n_a(t)$.

The broadening due to the charge transfer instability in both the absorption and the photoelectron emission may occur when $\eta^2 > \pi\Gamma$ as seen in Fig.1. Since the broadening is related to the presence of two minima in the adiabatic potential curve in the static theory, it is obtained only if the higher order terms, at least η^4, of the plasmon coupling is taken into account. Therefore, the perturbation theory taking into account the terms up to η^2 cannot give the broadening.

4 Comparison with Experiments

It is interesting to compare our theoretical results with the experiments of Cunningham et al[8,10] and Gibbs et al[9] reporting persistence and non-persistence of the atomic exciton absorption lines of rare gas adsorbates on metals, when the coverage is lowered below unity, according to the sign of

$\tilde{\varepsilon}_a(0)$ ($\equiv \phi - I^*$) being negative and positive, respectively: ϕ is the workfunction of the substrate and I^* is the ionization potential of the adsorbate. Cunningham et al[8] first pointed out that this is due to the charge transfer instability. If the non-persistence of the exciton lines could be ascribed to the spectral broadening, this idea may be justified by our theory only when the non-persistence and persistence condition is changed from $\tilde{\varepsilon}_a(0) \gtrless 0$ to $\tilde{\varepsilon}_a(0) \gtrless -\eta^2/2$.

The recent interpretation[9,10] of the experiments, however, is different from the first one,in that the optical transition to the neutral adsorbate excited state is considered to be observable, even when the charge transfered state is lower than the neutral one in energy. Such a neutral excited state is assigned to the metastable state obtained in our static theory[4], for example in the spinless model Hamiltonian system, to the state of minimum 2 in the region of $-\eta^2/2 < \tilde{\varepsilon}_a(0) < 0$ ($\phi < I^* < \phi + \eta^2/2$) and $\pi\Gamma < \eta^2$ of Fig.1. The observability is due to a long lifetime of the metastable state. In our dynamical theory, the broadening across the critical line may not be appreciable if Γ is very small. The optical transition to the metastable state is characterized by loss of the edge singularity as pointed out by Flynn's group[9,10] and seen in our result of Fig.3(b). Inconsistency[11] between the optical absorption and the electron energy-loss spectral widths would come from a difference in the methods of detecting the transition as already discussed.[9,10] According to our results, the photoelectron emission lines are expected to be sharp if $\tilde{\varepsilon}_a(0) > -\eta^2/2$ ($I^* < \phi + \eta^2/2$). This agrees with the photoelectron emission experiment.[11]

5 Luminescence from Adsorbates

Finally we would like to point out that the relative positions of the neutral and the charge-transferred excited states could be determined by observing e the luminescence from the adsorbate.

A dynamical calculation[12] of the luminescence spectral shapes similar to that for the absorption and the photoemission has been performed, under the assumption that the radiative lifetime is much longer than the time constant of the charge transfer: this assumption seems realistic in many systems. The calculated results are shown in Fig.5. Figure 5(a) corresponds to the case where the lowest excited state is neutral and Fig.5(b) to the opposite case.

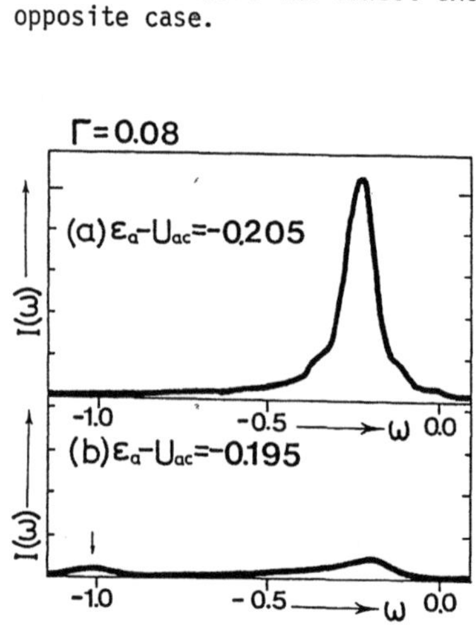

Fig.5 Calculated line shapes of the luminescence for η^2=0.4.

References

1 O. Gunnarsson and K. Schönhammer, Solid State Commun. 26, 147 (1978)
2 J.W. Gadzuk, Surf. Sci. 86, 516 (1979)
3 B. Gumhalter, J. Phys. (Paris) 38, 1117 (1977)
4 K. Shinjo, S. Sugano, and T. Sasada, Phys. Rev. B 28, 5570 (1983)
5 K. Shinjo and S. Sugano, Phys. Rev. B 30, 604 (1984)
6 G.D. Mahan, Phys. Rev. 163, 612 (1967)
7 P. Noziéres and C.T. de Domminicis, Phys. Rev. 178, 1097 (1969)
8 J.E. Cunningham, D.K. Greenlaw, and C.P. Flynn, Phys. Rev. B 22, 717 (1980)
9 D. Gibbs, J.E. Cunningham, and C.P. Flynn, Phys. Rev. B 29, 5292 (1984)
10 J.E. Cunningham, D. Gibbs, and C.P. Flynn, Phys. Rev. B 29, 5304 (1984)
11 J.E. Demuth, Ph. Avouris, and S. Schmeisser, Phys. Rev. Letters 50, 600 (1983)
12 K. Shinjo and S. Sugano, J. of Luminescence (to be published)

Optical Spectra of Adsorbates on Metal Surfaces

H. Ueba

Department of Electronics, Faculty of Engineering, Toyama University
Takaoka, Toyama 933, Japan

A. Yoshimori

Department of Material Physics, Faculty of Engineering Science
Osaka University, Toyonaka, Osaka 560, Japan

1. Introduction

A substantial progress has been made in spectroscopic studies of adsorbates on metal surfaces. Photoelectron spectra of core levels for adsorbates are certainly the central subject of the theoretical effort in which a main concern is placed on the effect of the screening of the core hole, both by charge transfer and by coupling to surface plasmons [1,2]. In addition to the shift in the main peak and the appearance of surface plasmon satellites, the characteristic feature of the core spectra includes the so-called "Fermi-edge singularity [3] , caused by the excitation of electron-hole pairs, since the relaxation shift of the adsorbate valence level upon a creation of the core hole gives rise to a localized final state perturbation.

Recently optical absorption spectra of adsorbates have also attracted the theoretical interest [4-6], in conjunction with the observation of a bimodal behavior of rare gas atoms on metal surfaces [7]. When one describes the valence state of adsorbates within the Newns-Anderson model, its optical spectra are essentially identical to that of a core electron in metal with an incomplete shell [8], which exhibit the singular absorption edge depending on the position of the localized state relative to the Fermi level. The purpose of this paper is to study the analytic features of the absorption spectra of adsorbates with a main interest on how the singular absorption edge is modified by the adsorbate coupling to the surface plasmons.

2. Theory

We consider the following spinless model Hamiltonian which has been employed in the studies of the above mentioned optical processes of adsorbates:

$$H = \varepsilon_c n_c + \varepsilon_a n_a + \sum_k \varepsilon_k n_k + \omega_p b_p^+ b_p + V\sum_k (c_a^+ c_k + h.c.) + \lambda(1-n_a-n_c)(b+b^+). \quad (1)$$

It is assumed here that in the initial state (n_c=1) of the optical transition, the energy of the unoccupied adsorbate level is high enough above the Fermi level ε_f so that both the mixing V with the metal electrons and the coupling λ between the net charge of the adsorbate and the surface plasmon with frequency ω_p are treated as the final state interaction. In the final state (n_c= 0), the unoccupied level of the excited adsorbate is lowered down to ε_a above ε_f due to the potential of a core hole left behind.

The line shape of the absorption spectra is calculated by taking the imaginary part of the adsorbate Green's function,

$$G_a(z) = \langle i|c_a G(z) c_a^+|i\rangle \ , \quad z = \omega - i\eta \ (\eta \to +0) \quad (2)$$

where $|i\rangle$ represents the eigenstate with the energy E_i of the initial state and $G(z)=(z-H_f+E_i)^{-1}$ and the final state Hamiltonian H_f is, hereafter, rewritten in terms of displaced plasmon operator ($b \to b-\lambda/\omega_p$). The photon energy ω is measured from $-\varepsilon_c$. Since we are interested in the Fermi-edge singularity, appropriate resolvent expansion of (2) with respect to both V and λ enables us to obtain

$$G_a(z) = \hat{g}_a(z) + \{V\hat{g}_a(z)\}^2 \sum_{kk'}^{>>} G_{kk'}(z) , \tag{3}$$

where $G_{kk'}(z)=\langle i|c_k G(z) c_{k'}^{\dagger}|i\rangle$ is composed of a collection of terms which give rise to the logarithmic divergence at $\omega \sim \varepsilon_f$, whereas the rest of the non-singular contribution is taken into account in $\hat{g}_a$. Figure 1 shows the graphical representation of $\hat{g}_a$ and $G_{kk'}$. By inspecting the structure of $G_{kk'}$, we found that the free propagator g_k of the metal electron is renormalized into

$$\hat{g}_k(z) = \left[g_k^{-1}(z) - V^2 \sum_{k'}^{<} \{g_{kk'a}^{-1}(z) - \lambda^2 g_{kk'ap}^{-1}(z)\}^{-1} \right]^{-1} , \tag{4}$$

where

$$g_k(z) = (z + v - \varepsilon_k)^{-1} , \quad g_{kk'a}(z) = (z - \varepsilon_k + \varepsilon_{k'} - \varepsilon_a - v)^{-1} ,$$

$$g_{kk'ap}(z) = g_{kk'a}(z - \omega_p) ,$$

here the summations $\sum_k^{><}$ extend over the states satisfying $\varepsilon_k >, < \varepsilon_f$ and $v=\lambda^2/\omega_p$ represents the static image potential. The explicit form of $\hat{g}_a$ is also given by

$$\hat{g}_a(z) = \left[g_a^{-1}(z) - \lambda^2 \{g_{ap}^{-1}(z) - V^2 \sum_k^{>} g_{kp}^{-1}(z)\}^{-1} \right]^{-1} , \tag{5}$$

where

$$g_a(z) = (z - \varepsilon_a - v)^{-1} , \quad g_{ap}(z) = g_a(z - \omega_p) , \quad g_{kp}(z) = g_k(z - \omega_p) .$$

In terms of $\hat{g}_a$ and $\hat{g}_k$, $G_{kk'}$ is then expressed as

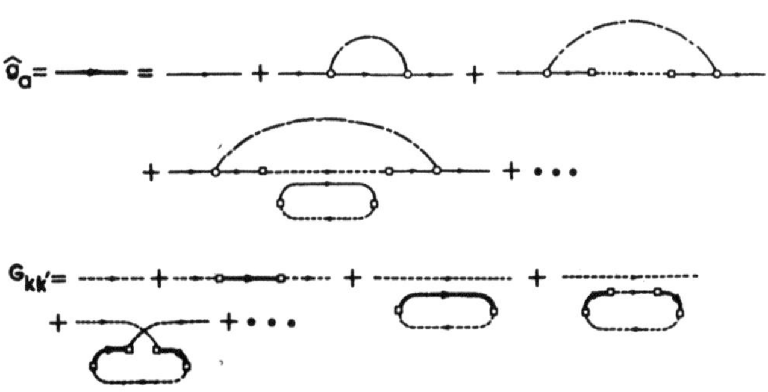

Fig. 1 Graphical representation of $\hat{g}_a$ and $G_{kk'}$. The solid (dashed) and dot-dashed lines represent propagators of adsorbate (metal) electrons and surface plasmons, respectively. The vertices □ and ○ stand for the mixing interaction V and the surface plasmon coupling λ. The higher order diagrams (• • •) are composed of a combination of these illustrated here.

$$G_{kk'}(z) = \hat{g}_k \delta_{kk'} + V^2 \hat{g}_k(z)\hat{g}_a(z)\hat{g}_{k'}(z)$$

$$+ V^4 \hat{g}_k(z)\hat{g}_a(z) \overset{<>}{\sum_{k''}\sum_{k'''}} \hat{g}_{k'k''k'''}(z)\hat{g}_a(z)\hat{g}_{k'}(z)\delta_{kk'} + \cdots \quad , \tag{6}$$

where $g_{k'k''k'''}$ is a renormalized one of $g_{k'k''k'''}(z)=(z-\varepsilon_{k'}+\varepsilon_{k''}-\varepsilon_{k'''})^{-1}$ in the spirit of (4). We introduce a rectangular density of states ρ with a half-band width D for metal electrons so that k-summation of (6) is completed as

$$G_m(z) = \overset{>>}{\sum_{kk'}} G_{kk'}(z) = \rho\{L(\omega) - gL^2(\omega) + \frac{g^2}{2}L^2(\omega) + \cdots\}, \tag{7}$$

where

$$L(\omega) = \log|(\omega - \tilde{\varepsilon}_f)/D| + i\pi\theta(\omega - \tilde{\varepsilon}_f) \ , \tag{8}$$

$$g = -\rho V^2 \hat{g}_a(\tilde{\varepsilon}_f) \ , \tag{9}$$

and $\theta(\omega)$ the usual step function. The threshold energy $\tilde{\varepsilon}_f$ is given by the solution of

$$\tilde{\varepsilon}_f = \varepsilon_f - v + \rho V^2 \log\left|\frac{\hat{g}_{ap}^{-1}(\tilde{\varepsilon}_f)}{\hat{g}_{ap}^{-1}(\tilde{\varepsilon}_f)-D}\right| \quad , \tag{10}$$

where

$$\hat{g}_{ap}^{-1}(\tilde{\varepsilon}_f) = (\tilde{\varepsilon}_f - \varepsilon_a - v - \frac{\lambda^2}{\tilde{\varepsilon}_f - \varepsilon_a - v - \omega_p}) \ . \tag{11}$$

Since $\hat{g}_a(z)$ in (3) has no imaginary part when $\omega \to \tilde{\varepsilon}_f+0$, the line shape near the threshold is given by

$$F(\omega) = \pi^{-1}\rho V^2 \hat{g}_a^2(\omega) G_m^I(\omega) \ , \tag{12}$$

where G_m^I, the imaginary part of (7) is given by

$$G_m^I(\omega) = \pi\{1 + (-2g + g^2)\log(\omega - \tilde{\varepsilon}_f)/D + \cdots\}\theta(\omega - \tilde{\varepsilon}_f) \ , \tag{13}$$

$$= \pi\{(\omega - \tilde{\varepsilon}_f)/D\}^{\alpha}\theta(\omega - \tilde{\varepsilon}_f) \ , \tag{14}$$

within the next divergent approximation and $\alpha=-2g+g^2$ represent the power of divergence, which exhibits the same structure as obtained in [8]. Consequently the adsorbate coupling to the surface plasmons results in the shift of the absorption threshold $\tilde{\varepsilon}_f$ and the modification of g from $g_0=\rho V^2/(\varepsilon_a-\varepsilon_f)$. In the limit $\omega_p>>\varepsilon_a-\varepsilon_f$ and ρV^2, we obtain the approximated expression of g from (9) as

$$g = \frac{g_0}{1 + \{1 + \frac{\omega_p}{\varepsilon_a - \varepsilon_f} + g_0 \log\frac{\omega_p}{\omega_p + D}\}S} \quad , \tag{15}$$

where $S=(\lambda/\omega_p)^2$. The intensity of the threshold singularity is therefore quenched in the presence of the adsorbate coupling to the surface plasmons.

Let us turn our attention to the spectra in the neighbourhood $\omega \sim \varepsilon_a$. In the perturbation expansion of (2), we take into account the diagrams of series connections of the adsorbate and the metal electron propagators supplemented with the surface plasmons, i.e., we obtain

$$G_a(z) = \left[g_a^{-1}(z) - \sigma(z) - \lambda^2 \{ g_{ap}^{-1}(z) - \sigma(z - \omega_p) \}^{-1} \right]^{-1} , \tag{16}$$

where

$$\sigma(z) = V^2 \sum_k^> \frac{1}{z + v - \varepsilon_k} = \rho V^2 \left[\log \left| \frac{\tilde{\omega}}{\tilde{\omega} - D} \right| + i\pi \{ \theta(\tilde{\omega}) - \theta(\tilde{\omega} - D) \} \right] , \tag{17}$$

$$\tilde{\omega} = \omega + v - \varepsilon_f ,$$

gives the energy-shift and the level broadening of the adsorbate state. Equation (12) and the imaginary part of (16) now complete the absorption spectra in the case $\varepsilon_a > \varepsilon_F$.

3. Results and Discussion

In what follows the results of numerical calculation are presented, where we take D as energy unit and ε_f as origin. The power α, and the position $\tilde{\varepsilon}_f$ of the Fermi-edge singularity are plotted in Fig. 2 as a function of λ, where ε_a=0.3, ω_p=0.6 and ρV^2=0.04, 0.1. It is found that α increases in negative sign with λ in accordance with (15). It is also observed in Fig. 2 that $\tilde{\varepsilon}_f$ shifts to lower energy side with λ and the magnitude of this shift is roughly given by $v+\rho V^2$.

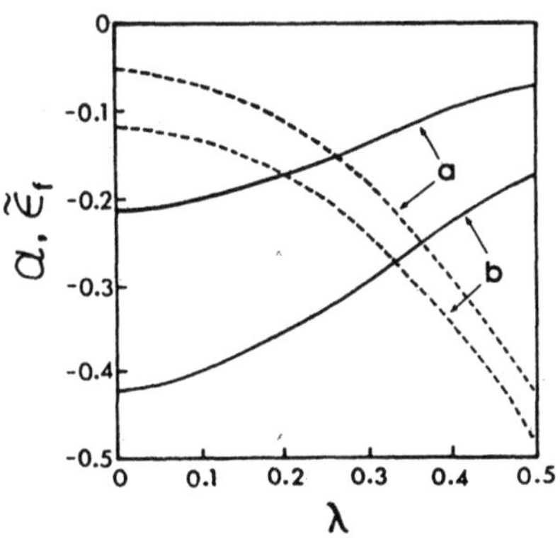

Fig. 2 The power α and the threshold energy ε_f of the divergence as a function of λ, where ε_a=0.3, ω_p= 0.6, and ρV^2=0.04(a), 0.1(b).

Examples of the optical absorption spectra are depicted for ε_a=0.3 and ω_p= 0.6, where ρV^2 and λ are the parameters. For λ=0.0, the spectrum (Fig. 3c) is characterized by the absorption threshold at $\tilde{\varepsilon}_f$ and a main peak with the width $\pi\rho V^2$ at $\omega \sim \varepsilon_a$. Further increase in ρV^2 then enhances the intensity at the threshold and causes more broadened peak as seen from Figs. 3a and 3b. For finite λ, the spectrum (Fig. 3c) exhibits characteristic changes into the spectrum (Fig. 3d), where the absorption threshold shifts to lower energy side by the order of v and its intensity is remarkably diminished. Another change includes a narrowing of the main peak accompanied by the appearance of the surface plasmon side-band with a weight of about S. It is mentioned here that

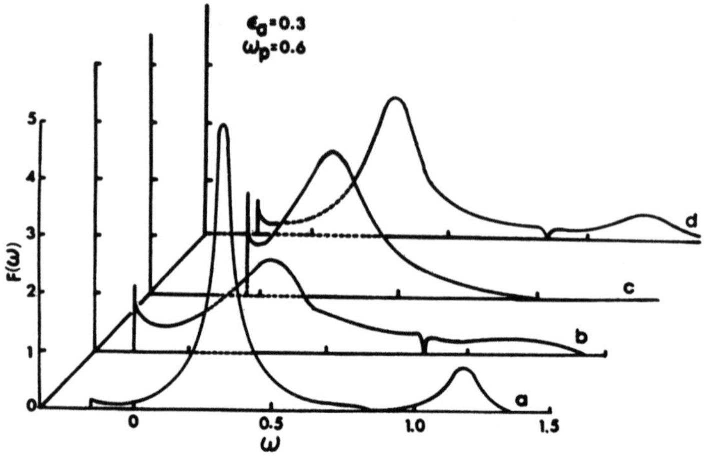

Fig. 3 Optical absorption spectra calculated for ε_a=0.3, ω_p=0.6, and $(\rho V^2,\lambda)$= a: (0.02,0.3), b: (0.06,0.3), c: (0.04,0.0), d: (0.04,0.3).

no antiresonance appears in the neighbourhood $\omega\sim\varepsilon_a$ since we have not considered a direct excitation of core electron into the unoccupied states of the metal.

It is also important to note that the absorption spectra also exhibit another singularity at the position spanned by ω_p from $\tilde{\varepsilon}_f$. Namely, the third diagram of $\hat{g}_a$ in Fig. 1, given by

$$\lambda^2\rho V^2\hat{g}_a^2(z)\hat{g}_{ap}^2(z)\left[\log\left|\frac{\hat{\omega}}{\hat{\omega}-D}\right| + i\pi\{\theta(\hat{\omega}) - \theta(\hat{\omega}-D)\}\right], \tag{18}$$

$$\hat{\omega} = \omega + V - \varepsilon_f - \omega_p ,$$

becomes the simplest unit responsible for the singularity at $\omega\sim\varepsilon_f-V+\omega_p$. It is found that the intensity of this singularity is substantially quenched by the order of S compared to that at $\tilde{\varepsilon}_f$, thereby being buried in the spectra of the main peak for small ρV^2. As ρV^2 increases, however, such a second singularity will appear in the broadened continuum spectra.

Although the present analysis is restricted to the case $\varepsilon_a>\varepsilon_f$, a similar treatment is also extended to the case $\varepsilon_a<\varepsilon_f$, in which the singular absorption edge appears at $\tilde{\varepsilon}_a$ with a tail growing on its higher energy side with ρV^2 [8]. But no significant effect of the surface plasmon would occur, since the adsorbate is neutralized by the screening charge. On increasing ε_a above ε_f, the absorption threshold at $\tilde{\varepsilon}_a$ is replaced by that at $\tilde{\varepsilon}_f$ accompanied by a peak at ε_a and those characteristic structures upon change of ρV^2 and λ have been studied in detail.

Finally, it should be noted that the present approach is no longer valid when ε_a is close to ε_f. The expression of g given by (9) is valid for small $\Delta/(\varepsilon_a-\varepsilon_f)$, $\Delta=\pi\rho V^2$. If this is large, g is expressed in terms of the phase-shift at the Fermi level, $\delta = \tan^{-1}(\Delta/(\varepsilon_a-\varepsilon_f))$ as $g=\delta/\pi$ [8] . On the other hand, even for the general interacting electrons, δ/π is proved to be number of localized electrons to screen the excess charge of the impurity [9] . If this impurity can be replaced by the adsorbate, δ/π should be the screening charge including the effect of the image charge, namely that of the coupling to the surface plasmon. The proof [10] suggests that the singularity in the present case is also given as $\alpha = -2(\delta/\pi)+2(\delta/\pi)^2$ (spin to be taken into account) in terms of the phase-shift including the effect of the surface plasmon. The present result may be consistent to this picture within our approximation.

References

1 K. Schönhammer and O. Gunnarsson: Solid State Commun. 26, 147 (1978), Z. Phys. B30, 297 (1978)
2 B. Gumhalter and D. M. Newns: Phys. Letters 57A, 423 (1976) B. Gumhalter: J. Phys. (Paris) 38, 1117 (1979)
3 P. Nozieres and C. T. De Dominicis: Phys. Rev. 178, 1097 (1969)
4 B. N. J. Persson and Ph. Avouris: J. Chem. Phys. 79, 5156 (1983).
5 K. Shinjo and S. Sugano: Phys. Rev. B30, 604 (1984)
6 M. Tsukada and W. Brenig: preprint
7 J. E. Cunningham, D. K. Greenlaw, and C. P. Flynn: Phys. Rev. B22, 717 (1980)
8 A. Kotani and Y. Toyozawa: J. Phys. Soc. Jpn. 35, 1073 (1973)
9 J. S. Langer and V. Ambegaokar: Phys. Rev. 121, 1091 (1961)
10 K. Yamada and K. Yosida: Prog. Theor. Phys. 62, 363 (1979), 68, 1504 (1982)

Part III

Ordering

Rare Gas Monolayers on Crystal Surfaces. Impurity Effects

J. Villain

Institut für Festkörperforschung der Kernforschungsanlage Jülich GmbH
Postfach 1913, D-5170 Jülich 1, Fed, Rep. of Germany

M.B. Gordon

DRF/DN, Centre d'Etudes Nucléaires, 85 X, F-38041 Grenoble Cedex, France

As far as qualitative aspects are concerned, the phenomena to be discussed in the present review are not special to a particular physical system. However, because of the diversity of surface Physics, it is of interest to consider a special case, namely Krypton monolayers on graphite.

Why graphite? Because it is easily cleaved, and has a high symmetry. Certain halides have similar properties, but have not been so much investigated. Furthermore, graphite fibres have some technological application, and rare gas adsorption may be used as a test of their quality.

The interest in rare gases as adsorbates is obvious: they are simple. No question arises about molecular orientation, and the energy depends on a restricted number of parameters which are fairly well known. Of course, chemisorption on metals is technologically more interesting, but microscopic aspects are not so well understood, and it would be reckless to deal with collective properties, as we intend to do here. Or, at least, any precise comparison with experiment would be excluded.

Why choose Krypton rather than any other rare gas? Krypton monolayers on graphite have the remarkable property of undergoing a commensurate-incommensurate (C-I) transition, first discovered by Thomy and Duval in 1967 /1/ and investigated by many experimentalists /2,3/. In other words a Krypton film at appropriate temperature T and pressure P has a structure commensurate (C) with its substrate (Fig. 1A) in which it chooses one of 3 allowed sublattices. At other pressures the adsorbate becomes an incommensurate, two-dimensional solid (Fig. 1B). At higher temperature melting occurs, but the liquid phase will not be considered because there is no quantitative, analytical theory of melting and liquid state. Solid monolayers of Ne and Ar are always incommensurate because the atomic diameter is too different from the atomic distance in any simple commensurate phase. This distance is 4.26Å in the "$\sqrt{3}\times\sqrt{3}$" structure represented by Fig. 1A.

Why is the C-I transition so interesting? First of all, it is a fairly common property of adsorbed films, which has practical consequences on other properties, e.g. transport. It is also of fundamental interest, as is any phase-transition or, more generally, any non-linear phenomenon. Phase-transitions are indeed a severe test of the ability of statistical mechanics to determine qualitative, macroscopic properties, generally described in such cases by "critical exponents". But, in the particular case of Kr monolayers on graphite, the C-I transition is also a test of the microscopic model, because the theoretical analysis is sufficiently simple to be carried out quantitatively, without extracting qualitative, long wavelength features such as critical exponents.

In Sections 2 and 3, the standard, microscopic description of solid, rare gas monolayers on graphite is recalled and compared with experiment. In spite of remarkable successe, there is a serious flaw: namely, the commensurate structure of Kr films is found to be unstable, at least at temperature T=0, as will be seen in Section 4. Furthermore, the predicted critical behaviour of the "misfit" (the shift of the Bragg peak from the commensurate position)

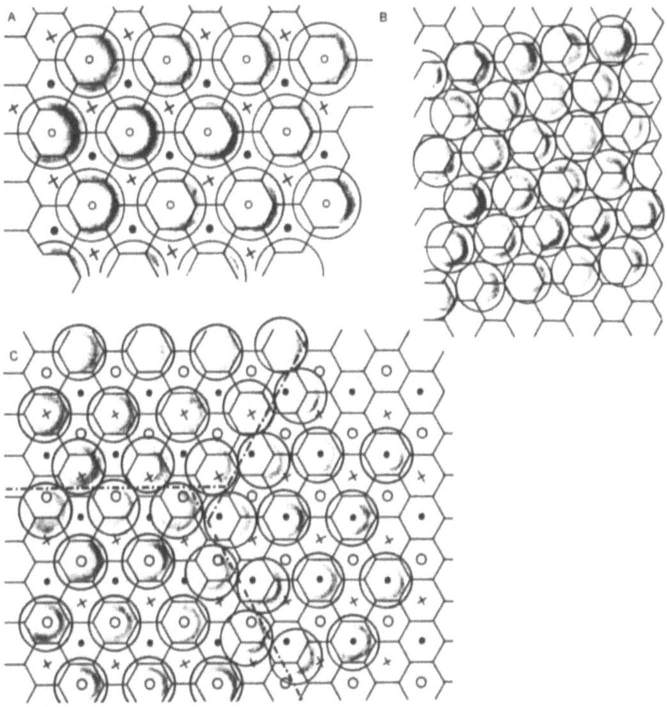

Figure 1. Solid structures of a krypton monolayer adsorbed on graphite. The circles denote rare gas atoms. The easy adsorption sites are the centres of the hexagons and the 3 sublattices are indicated by crosses, full dots and empty dots. A. Commensurate (C) phase. B. Incommensurate phase.C. The incommensurate phase near the C-I transition, showing 3 domains and 3 domain walls.

is not observed experimentally. This point is dealt with in Section 5. In Sections 6 and 7 the effect of frozen impurities or other defects is considered, and it is argued that they can account for some of the disagreements between theory and experiments.

2. Microscopic description of rare gas monolayers on graphite

The adatoms occupy positions $\vec{R}$, which are usually assumed to lie in a well-defined plane. They are subject to the substrate potential, which for an ideal surface is periodic. It can therefore be expanded in Fourier series. The adatoms will be treated classically, and pair interactions will be postulated. The Hamiltonian is therefore [4,5,6]

$$H = \sum_{<RR'>} U(\vec{R}-\vec{R}') - \sum_{GR} V_G \exp(i\vec{G}\cdot\vec{R}) \tag{2.1}$$

where $\vec{G}$ is a reciprocal lattice vector of the substrate lattice and $U(r)$ is not too different from a 6-12 Lennard-Jones potential

$$U(r) = 4\varepsilon[(\sigma/r)^{12}-(\sigma/r)^{6}] \tag{2.2}$$

It is correct to neglect the lattice strain because the covalent interaction of C atoms is much larger than the Van der Waals interaction with the adatoms. This assertion is confirmed by unpublished calculations based on the formulae of Lau and Kohn (7).

The summation on $\vec{G}$ in (2.1) can be restricted to the 6 smallest reciprocal lattice vectors, which form a regular star. Symmetry obliges all 6 V_G to be equal and real, $V_G=V=V^*$. Thus, there are only 3 real parameters ε,σ and V. The first two are known, $\varepsilon/K_B=158K$ and $\sigma=3.6Å$ for Kr atoms. V can be calculated assuming Lennard-Jones interactions between Carbon and rare gas and evaluating the corresponding constants ε',σ' as indicated by Steele /4/. In this way, Novaco and Mc Tague could predict quantitatively /5/ the rotation of Argon films with respect to the substrate (Fig. 1B). This rotation is due to the lower energy of shear strains compared to compression.

The rotation of Neon films was also observed by Calisti and Suzanne /8/ and the amplitude was that predicted by Novaco and Mc Tague several years before. Xe films are not rotated [9] and Kr films may or may not be rotated depending upon the temperature [10], presumably because they are too close to the commensurate state. In this case the approximation used by Novaco and Mc Tague [5] is no longer correct. Novaco and Mc Tague just linearized the second term of (2.1), and this is not correct near the C-I transition. What can then be done to take the non-linearity properly into account? This will be explained in the next Section.

3. Domains and walls.

In the simplest approach, the first term of (i) is treated in the harmonic approximation and the temperature is assumed to be zero. Neglecting a constant formula (2.1) reads

$$H = \sum_{\alpha,\gamma} \sum_{R,R'} g^{\alpha\gamma}_{RR'}(U^\alpha_R-U^\alpha_{R'})(U^\gamma_R-U^\gamma_{R'})-\sum_{G,R} V_G \exp(i\vec{G}\cdot\vec{R}) \qquad (3.1)$$

where α,γ = n,y are the coordinates and $\vec{U}_R$ is the displacement of adatom $\vec{R}$ with respect to the sites of a regular triangular lattice which may be chosen in various ways. A natural choice [5] far from the C-1 transition is the equilibrium structure for V_G=o. Another possibility appropriate close to the C-1 transition [6] is to choose one of the 3 commensurate sublattices (Fig.1A).

In the commensurate phase the second term of (3.1) is minimum. When the pressure is slightly higher than the transition pressure, the incommensurate phase looks like an array of commensurate domains separated by walls (Fig.1C). There is no strong experimental support for this statement, which mainly relies on a one-dimensional model introduced a long time ago by Frenkel and Kontorova [11] and studied in more detail by Frank and van der Merwe [12]. Application to adsorbed monolayers was suggested by Venables [13] and extensively utilized by the authors [6,14].

The orientation and structure of domain walls at T=o according to Refces 12 and 14 is displayed by Figure 1C. Analytic calculations [15] show that this orientation is stable with respect to small rotations. This does not rule out the possibility of completely different structures as speculated by Huse and Fisher [16] and Kardar and Berker [17]. However, the structure of Fig. 1C is confirmed by computer simulations of Abraham et al [18].

The wall concept is extremely useful at low temperature. It is indeed easy to calculate the energy of a wall, and this energy is related, through elementary thermodynamics, to the pressure P_c at which the C-1 transition occurs [6]. A more complicated problem is the calculation [19] of the wall-crossing energy Λ, the sign of which determines the order of the transition [20]. Such calculations are generally done in the "continuum limit" which corresponds to a weak substrate potential V_G. In the case Λ<o the C-I transition is first order. In the case Λ>o it is continuous, but one does not expect a direct transition from Fig.1A to Fig.1C. A simple argument, outlined at the end of this Section, shows [20] that there is an intermediate, "striped" phase in which walls are parallel (Fig. 2A).

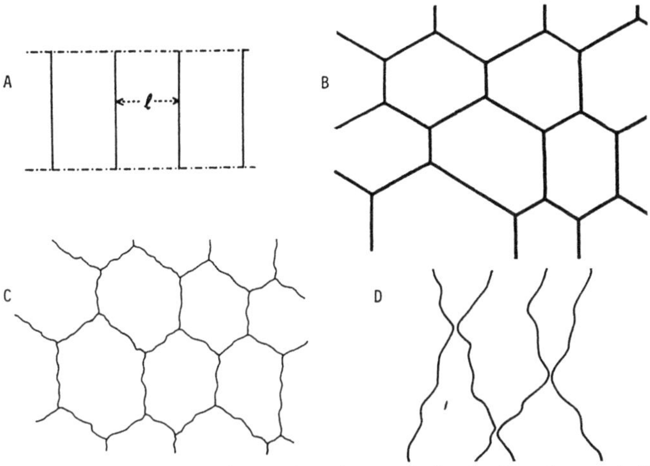

Figure 2. Wall configuration in a striped structure at T=o (A) in the hexagonally symmetric phase at low (B) and higher temperature (C), and in the striped phase at T≠o (D). The meanders in (C) and (D) can also be produced by impurities.

This (exact) argument is in contradiction with experiment according to which there is a second-order transition without striped phase. The same result was reached in finite temperature computer simulations by Abraham et al [18]. However, at T=o numerical calculations by Shiba [21] revealed a striped phase stable in an extremely narrow pressure range. All these results are consistent with a very small positive value of the wall-crossing energy Λ.

The argument of ref.20 relies on the form of the energy in the striped phase [6,12]

$$W_S = \frac{2N}{3\ell}\left[\xi_c - \xi + 4\xi_c \exp(\kappa\ell)\right] \tag{3.2}$$

while in the hexagonally symmetric phase [6]

$$W_H = \frac{4}{3}\frac{N}{\ell}\left[\xi_c - \xi + \xi_1 \ell^{\alpha} \exp(-\kappa'\ell) + \Lambda/\ell\right] \tag{3.3}$$

Here ξ is the chemical potential, ξ_c, ξ_1, Λ, α, κ, κ' are positive, real constants and ℓ is the distance between parallel walls. The exponential terms are indirect interactions mediated by the strained adsorbate. The energy of the commensurate state is taken as the origin. Minimisation of (3.2) and (3.3) leads to the above stated results.

4. Thermal effects.

What happens when the temperature T becomes finite? At sufficiently high temperature, the monolayer becomes liquid. It was even shown [22] that the hexagonally symmetric, weakly incommensurate solid phase is unstable at any T≠o with respect to formation of free dislocations. In other words, it melts. Actually, this effect can be disregarded at low temperature for two reasons: i) the density of dislocations is low. ii) as seen in Section 3 the nearly commensurate phase is not hexagonally symmetric, but striped. The striped structure is stable [22,23] if the number of sublattices (here, 3) is larger than $\sqrt{8}$.

Neglecting dislocations, the weakly incommensurate phase may be regarded as a slightly distorted array of domain walls (Fig. 2B and C). This hypothesis was extensively used by the authors [6,12]. It was confirmed by computer simulations of Abrahams et al [18]. The irregularities of the structure produce some entropy and this favours the incommensurate phase when T increases. This effect, however, is small. A more important effect, which acts in the same direction, is the thermal motion of the adatoms. In the commensurate phase, the last term of (2.1) is not really minimum at $T \neq 0$, so that the substrate potential is not so efficient in favouring the commensurate structure [6,21]. On the other hand, thermal expansion of Krypton (much larger than that of graphite) favours the commensurate state when the temperature increases. Thermal expansion is an anharmonic effect, not included in the approximation (3.1). Within approximation (3.1), the C-1 transition is expected to occur at a chemical potential $\xi_c(T)$ which decreases with increasing T [25,26,27,6]. Experimentally, $\xi_c(T)$ can easily be deduced from the corresponding pressure $P_c(T)$ measured by Larher [28] and by Fain et al [10]. It turns out that $\xi_c(T)$ has a maximum at about 80K. This is probably a result of the anharmonicity [6], the importance of which was first stressed by Venables et al. [13].

If one uses the parameters calculated by Steele [4] and used with success by Novaco and Mc Tague [5], it turns out that the commensurate phase should not be stable at all at zero temperature [29] when the "two-dimensional pressure" is positive, as it should be. The appearance of an incommensurate phase at higher temperature may possibly be explained by thermal expansion [27], but computer simulations [30,31,32] which take thermal expansion into account, reveal no commensurate phase at any temperature except when Steele's data [4] are conveniently modified [18]. Actually, the data used by Abraham et al [18] imply a substrate potential modulation twice as large as those of Steele. In our opinion, such a strong modulation is a priori not excluded.

It might result from a lower value of the Lennard-Jones parameter σ of the Kr-Carbon interaction, and this is the hypothesis of Ref.18. Also, it is not obvious that the Kr-C interaction is isotropic, and the strongly directed C-C interactions might produce a stronger modulation of the potential felt by the adatoms.

What is probably to be done in order to clarify the situation is the following: compute numerically (e.g. by molecular dynamics) the free energy of the commensurate phase and the free energy of the structure with a single wall. Subtraction yields the wall free energy. The wall-crossing energy Λ can be calculated in a similar way. From these data the phase diagram is easily obtained, while computer simulations may be subject to metastability effects [32]. It is worth mentioning that also in real experiments thermal equilibrium is not always reached at very low temperature.

5. Critical behaviour near the C-1 transition

The average distance ℓ between walls can be observed by Xray or electron diffraction since the Bragg peaks are shifted from the commensurate position by an amount proportional to $1/\ell$. In the case of a striped structure, Pokrovski and Talapov found [24] in the neighbourhood of the C-1 transition in two dimensions

$$1/\ell \simeq C(T)\sqrt{|\xi-\xi_c|} \tag{5.1}$$

The coefficient C(T) diverges at low temperature [33,34] as

$$C(T) \simeq \mathrm{Const}/T \tag{5.2}$$

For longer values of $|\xi-\xi_c|$ there is a crossover to a non-algebraic behaviour roughly described by the formula

$$|\xi-\xi_c| \simeq C'\ell \quad \exp(-\kappa\ell) \tag{5.3}$$

where $1/\kappa$ is the wall thickness and C' is a temperature-independent constant. Finally, for large values of $|\xi-\xi_c|$, the compressibility is roughly constant and $1/\ell$ varies roughly linearly with pressure or chemical potential, but in this region the wall picture makes no sense because the distance ℓ is of order the thickness $1/\kappa$.

Formula (5.3) results from the minimisation of the energy (3.2). Formula (5.1) results from an effective repulsion of entropic nature, since walls like to have more space available for their thermally induced meanders (Fig. 2D). Glachant et al [35] did observe formula (5.1) on Xenon monolayers on the [110] face of Cu. On this anisotropic surface, only striped structures, with a well-defined wall direction are allowed. The apparent agreement between experiment and theory is not quite convincing because the theoretical evaluation [33] of C(T) in (5.1) is about 10 times as large as the experimental result, and the theoretical domain of validity of (5.1) is much narrower than experimentally observed. Relations (5.2) and (5.3) could not be checked experimentally, possibly because the experiments on Cu[110] were not pushed to such a high accuracy as those on graphite.

What is expected instead of (5.1) and (5.3) in the case of hexagonal symmetry? For sufficiently large $|\xi-\xi_c|$, the behaviour is probably exponential as in (5.3), with possibly a different prefactor $\ell^{1+\alpha}$ and a different coefficient κ', as seen from (3.3). κ' is no longer the reciprocal wall thickness, and its knowledge requires the computation of the interaction between wall intersections. This is possible, but has not yet been done.

If one reproduces the type of argument which leads to (5.1) in anisotropic systems, one finds [6] a first-order transition at low enough temperature $T \neq 0$. Experimentally, Nielsen et al [36] did find a first-order transition when they diluted Kr by deuterium, but this may be explained by other mechanisms [17]. In pure Kr films the experimental observation [3,10] has no relation with theory! One finds

$$1/\ell \sim |\xi-\xi_c|^{1/3} \tag{5.4}$$

Although the exponent 1/3 accounts fairly well for experimental observations, it does not seem to have any theoretical basis. The law (5.4) may accidentally fit a more complicated behaviour resulting, for instance, from a crossover.

A general conclusion which may be drawn from this and the previous Sections is that the agreement between theory and experiment is not always ideal. A usual attitude in such a situation is to wonder whether impurities or other defects may contribute to this discrepancy. The effect of impurities will be considered in the next Section.

6. Effect of quenched impurities on the commensurate-incommensurate transition

Impurities have a big effect on C-I transitions and this effect is especially large in the two-dimensional case, which is of interest here. The basic mechanism is that an impurity sitting at a given site wants this site to be in one of the 3 allowed types of domains (the number 3 specifically applies to rare gas monolayers on graphite).

For a given domain shape of linear size ℓ, the impurity-induced energy fluctuation is, in D=2 dimensions, proportional to $\ell^{D/2}=\ell$ according to the central limit theorem. The domain shape adjusts in order to lower its energy, and the result of this adjustment is always an energy gain. In other words, the energy correction due to impurities is always negative. For rigid walls in the hexagonally-symmetric phase [37] this correction is proportional to ℓ for each domain. Since the number of domains is proportional to N/ℓ^2, where N is the area in appropriate units, the energy correction is $-NH/\ell$ where H is a constant which depends on the nature of the impurities. This correction is absent in the striped phase, so that impurities favour the hexagonal phase (3.7). However, for large values of ℓ, this argument is not complete and one must consider that walls are not rigid, but make meanders under the effect of impurities. The energy gain in two dimensions [38,39,40,41] is approximately

$$\delta W_1 \approx -\frac{1}{2}(N/\ell)H^{4/3}\, g^{-1/3}\ell n\, \ell-(N/\ell)H \tag{6.1}$$

for the hexagonally symmetric phase, and

$$\delta W_2 \approx -\frac{1}{2}(N/\ell)H^{4/3}\, g^{-1/3}\left[\ell n\, \ell + \frac{2}{3}\, \ell n(g/H)\right] \tag{6.2}$$

for the striped phase. g is the surface tension of the walls. Once more one finds that the hexagonally symmetric phase is stabilized by impurities. Another effect is that the commensurate phase is, in principle, never stable for an infinite, two-dimensional system with quenched impurities. The reason is that the negative, logarithm term in (6.2) dominates all terms of (3.3).

The explicit calculation in the hexagonally-symmetric phase would require the knowledge of Λ and α in (3.3). In the striped phase there are not so many parameters. The quantity $1/\ell$, which characterizes the degree of incommensurability, may be determined by minisation of the sum of (3.2) and (6.1). It turns out [38] that, for some values of the random field, $1/\ell$ is accurately described (Fig. 3) by relation (5.4)! This is, of course, completely accidental. As seen from Fig.3, the incommensurability is unobservable at low pressure, so that, experimentally, the system looks commensurate.

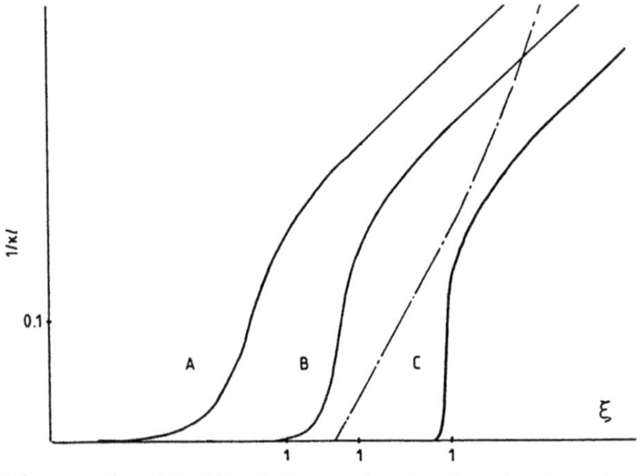

Figure 3. Misfit $1/\ell$ against chemical potential ξ in a striped structure with quenched impurities for D=2. Full curves ABC correspond to various values of H and the dashed line shows the cube of curve B, which looks like a straight line just above the transition, in agreement with (5.4).

If the temperature is not too high, the entropy effects responsible for the first-order transition predicted in [6] are dominated by disorder effects and the first-order transition is not observed. On the other hand, at high temperature, the calculation of ref. [6] is not valid.

The problem dealt with in the present Section is analogous to the very controversial Ising model in a random field. We adopted the most commonly admitted standpoint [38,39,40,41,42,43,44] according to which the lower critical dimension D_ℓ is two. Indeed, the logarithms in (6.1) and (6.2) are typical of a system at its lower critical dimension. However, there are theories [45,46,47] which predict $D_\ell=3$, and more recent treatments [48] still support this result. Experiments on random field Ising, magnetic systems have been done in 2 an 3 dimensions [49,50,51]. Some of them [49,50] seem consistent with $D_\ell=2$, while the difficulties encountered by other authors can be explained by metastability [51,52,44].

To summarize this Section, impurity effects can explain the following features of the C-1 transition of Kr monolayers on graphite. i) it is always observed as continuous. ii) the striped phase was never found. iii) the empirical law (5.4).

References

[1] A. Thomy, J. Regnier, X. Duval, in Thermochimie, Colloques Internationaux du C.N.R.S. (C.N.R.S., Paris) 201, 511 (1972). See also Surface Sci. Rept. 1, 1 (1981)

[2] S.C. Fain, Jr., and M.D. Chinn, J. Physique 38, C4-99 (1977)

[3] D.E. Moncton, P.W. Stephens, R.J. Birgenau, P.M. Horn, G.S. Brown, Phys. Rev. Letters and refces therein 46 1533 (1981)

[4] W.A. Steele, Surface Sci. 36 317 (1973)

[5] A.D. Novaco, J.P. Mc Tague, Phys. Rev. Letters 38, 1286 (1977); J. Physique 38 C4-116 (1977), Phys. Rev. B19, 5299 (1979)

[6] J. Villain, M.B. Gordon, Surface Sci. 125, 1 (1983)

[7] K.H. Lau, W. Kohn, Surface Sci. 65, 607 (1977)

[8] S. Calisti, J. Suzanne, Surface Sci. 105, L255 (1981)

[9] P.S. Schabes-Retchkiman, J.A. Venables, Surface Sci. 105, 536 (1981)

[10] S.C. Fain, Jr., M.D. Chinn, R.D. Diehl, Phys. Rev. B21, 4170 (1980)

[11] Y.I. Frenkel, T. Kontorova, Zh. Eksp. Teor. Fiz. 8, 1340 (1938)

[12] F.C. Frank, J.H. van der Merwe, Proc. R. Soc. London 198, 205 (1949)

[13] J.A. Venables, P.S. Schabes-Retchkiman, J. Physique 38, C4-105 (1977)

[14] J. Villain, in Ordering in Strongly Fluctuating Condensed Matter Systems, Ed. T. Riste (Plenum, New York, 1980) p.221

[15] M.B. Gordon, J. Villain, J. Phys. C15, 1817 (1982)

[16] D.A. Huse, M.E. Fisher, Phys. Rev. Lett. 49, 793 (1982)

[17] M. Kardar, A.N. Berker, Phys. Rev. Lett. 48, 1552 (1982)

[18] F.F. Abraham, W.E. Rudge, D.J. Averbach, S.W. Koch, Phys. Rev. Lett. 52, 445 (1984)

[19] A.L. Talapov, Phys. Rev. B24, 6703 (1981)

[20] P. Bak, D. Mukamel, J. Villain and K. Wentowska, Phys. Rev. B19, 1610 (1979)

[21]H. Shiba, J. Phys. Soc. Japan 48, 211 (1980)
[22]S.N. Coppersmith, D.S. Fisher, B.I. Halperin, P.A. Lee and W.F. Brinkman, Phys. Rev. Letters 46, 549 (1981)
[23]F.D.M. Haldane, P. Bak, T. Bohr, Phys. Rev. B28, 2743 (1983)
[24]M. Schöbinger, S.W. Koch, Z. Physik B53, 233 (1983)
[25]V.L. Pokrovskii and A.L. Talapov, Phys. Rev. Letters 42, 65 (1979)
[26]V.L. Pokrovskii, A.L. Talapov "Theory of incommensurate crystals". Soviet Scientific Reviews, Supplement Series, Phys. Vol.1 (Harwood, Chur, Switzerland)
[27]V.E. Kravtsov, Solid State Commun. 38, 17 (1981)
[28]Y. Larher, J. Chem. Phys. 65, 2257 (1978)
[29]M.B. Gordon, J. Villain, to be published
[30]T.H. Spurling, J.E. Labe, Aust. J. Chem. 31, 465 (1978)
[31]J.S. Whitehouse, D. Nicholson, N.G. Parsonage, Molec. Phys. 49, 829 (1983)
[32]S.J. Knak Jensen, J. Phys. C16, 4509 (1983)
[33]F.D.M. Haldane and J. Villain, J. Physique 42, 1673 (1981)
[34]Y. Okwamoto, J. Phys. Soc. Japan 52, 942 (1983)
[35]A. Glachant, M. Jaubert, M. Bienfait and G. Boato, Surface Sci. 115, 219 (1981)
[36]M. Nielsen, J. Als-Nielsen, J. Bohr, J.P. Mc Tague, Phys. Rev. Lett. 47, 582 (1981)
[37]J. Villain, J. Physique Lett. 41, L267 (1980)
[38]T. Nattermann, J. Phys. C16, 4113 (1983)
[39]K. Binder, Z. Phys. B50, 343 (1983)
[40]G. Grinstein, S.K. Ma, Phys. Rev. B28, 2588 (1983)
[41]J. Villain, B. Séméria, F. Lancon, L. Billard, J. Phys. C16, 6153 (1983)
[42]D. Andelman, H. Orland, L.C.R. Wijewarhana, Phys. Rev. Lett. 52, 145 (1984)
[43]D.S. Fisher, J. Fröhlich, T. Spencer, J. Stat. Phys. 34, 863 (1984)
[44]R. Bruinsma, G. Aepli, Phys. Rev. Lett. 52, 1543 (1984)
[45]D. Mukamel, E. Pytte, Phys. Rev. B25, 4779 (1982)
[46]H.S. Kogon, D.J. Wallace, J. Phys. A14, L527 (1981)
[47]J.L. Cardy, Phys. Lett. 125B, 470 (1983)
[48]U. Krey, Regensburg preprint (1984)
[49]D.P. Belanger, A.R. King, V. Jaccarino, J.L. Cardy, Phys. Rev. B28, 2522 (1983)
[50]R.A. Cowley, R.J. Birgenau, G. Shirane, H. Yoshizawa: Multicritical Phenomena (Ed. R. Pynn, Plenum, New-York 1983)
[51]P.Z. Wong, J.W. Cable, Phys. Rev. B28, 5361 (1983)
[52]J. Villain, Phys. Rev. Lett. 52, 1543 (1984) and to be published in the proceedings of the workshop on elementary excitati-ns and fluctuations in magnetic systems at San Miniato (Ed. S. Lovesey, U. Balucani and V. Tognetti, Springer Verlag)

Addendum

In agreement with Section 4, it is pointed out in various articles (Gooding et al, Phys. Rev. B27, 7669, 1983, and Vidali and Cole, Phys. Rev. B27, 6436, 1984) as well as in recent preprints (Sander and Hartman, Schöbinger and Abraham) that the substrate corrugation is underestimated by a factor of about two in the case of Kr films graphite. This conclusion results from the experimental properties of the monolayer, but also from He scattering experiments (Boato et al Phys. Rev. B20, 3957, 1979 and Derry et al, Surf. Sci.

87, 629, 1979). It should be stressed that the agreement with experiment obtained by Novaco and Mc Tague [5] (see also Phys. Rev. B19, 5299, 1979) is not a test of the parameter V. Indeed, within second-order perturbation theory, the rotation angle depends only on the Poisson ratio $\tilde{\sigma}$. For two-body central forces between adatoms, $\tilde{\sigma}$ has a universal value according to the Cauchy rule. Thus, the success of the Novaco-Mc Tague theory just implies that the interaction between adatoms is pair-wise with a good approximation.

We are grateful to S. Koch, S. Fain and H. Shiba for their comments, on which this addendum is based.

The Lattice Gas Model of Surface Ordering

Junjiro Kanamori

Department of Physics, Faculty of Science, Osaka University
Toyonaka, Osaka 560, Japan

A summary of the method of geometrical inequalities which is capable of determining rigorously the ground states of the lattice gas model with competing interactions is presented. The results for the two-dimensional lattices are surveyed with the purpose of applying the model to the problems of ordering on the surface. The prediction of infinite series of the ground states in the case of the honeycomb lattice and the (7×7) reconstruction and related ones of the (111) surface of Si, Ge and their alloys are discussed in some detail as illustrative examples of the applications.

I. Introduction

The lattice gas is a useful model of ordering phenomena on the surface to elucidate their statistical mechanical aspect. As will be discussed below, it can cover not only the ordering of adsorbates but also some cases of surface reconstruction. With a relatively short range interaction, the model can exhibit a variety of the ground states. A rigorous analysis, however, is required to conclude the correspondence between an observed ordered state and the ground state of the model for assumed interactions. In some cases we need a careful analysis to infer the nature of interactions from the observed ordered pattern. In other cases it is not evident what model of interactions will yield the observed ordered pattern as the ground state. We shall discuss in this paper primarily the determination of the ground states of the model with competing interactions. The present paper will supplement previous discussions of surface-ordering with relatively simple lattice gas models reviewed by SELKE, BINDER and KINZEL [1], where the phases at finite temperatures are brought into focus rather than the ground states.

The ground states of the one dimensional lattice gas model can be determined rigorously by several methods including the transfer matrix one. The problem in higher dimensions, however, cannot be solved generally. We discuss in this paper the method of geometrical inequalities which was proposed in 1966 by the present author [2] and later developed in cooperation with KABURAGI [3, 4]. The method is the most powerful one so far known, in the sense that it determined the ground states in many cases where other methods failed. It makes use of inequalities which restrict the lowest possible values of linear combinations of the numbers of the nn, nnn, ··· pairs of particles for a given total number of particles on a lattice. When more-than-two particle interactions are introduced, the inequalities can be extended to include the numbers of various kinds of triplets, quartets, etc. The method cannot give a general solution, because it needs to derive the inequalities in individual cases. We have invented, however, a few general methods of deriving the inequalities which have proved to be effective in solving many problems [3, 4].

We give a general discussion on the ordering of particles in the lattice gas model and a summary of the method of geometrical inequalities in Section II. In section III we review briefly the analyses of the ground states of the two-dimensional lattice gas model. In section IV and V we discuss two recent applications. One is the honeycomb lattice where three kinds of infinite series of the ground states are found. [5] Another is the discussion of the (7×7) reconstruction and related ones of the (111) surface of Si, Ge and their alloys. [6] The final section (VI) concludes this paper.

II. Ground States and Geometrical Inequalities

We assume a Bravais lattice in the following discussion, for simplicity, though it can be generalized to non-Bravais lattice cases easily. The total number of lattice points is denoted by N and the total number of particles by Nx with x representing the density. We start with the discussion of the dependence on x of the lowest energy E_g of the lattice gas model with competing interactions. In the limit of infinite N the ground state energy per site, E_g/N is given as a function of x by a broken line having several inflection points where the x dependence changes and being dependent linearly on x inbetween (see Fig. 1). The ground state with x corresponding to an inflection point is given generally by a regularly ordered configuration of particles, though in some cases a finite or infinite number of particle configurations are degenerate in energy with it. Off the inflection point the ground state is generally a two-phase mixture of the ground states at adjacent inflection points; note that the boundary energy between the two phases can be assumed to be negligible in E_g/N in the limit of infinite N.

The characteristic densities corresponding to the inflection points are determined by the lattice geometry and relative magnitudes of the interactions. A simple example demonstrating these facts is given by the case where the interaction of the nn pairs of particles is strongly repulsive. We denote by p_1 the total number of the nn pairs in a given particle configuration. In the case of the triangular lattice we can prove the inequality given by

$$p_1 \geq \mathrm{Max}[0,\ N(3x-1),\ N(6x-3)], \qquad (1)$$

where the r.h.s. represents the maximum among those listed in the parenthesis. We discuss the derivation of inequalities such as (1) later in this section. The inequality (1) tells us that the maximum density under the condition $p_1 = 0$ is $x = 1/3$ provided that the particle configuration satisfying the equality at $x = 1/3$ can be found. With the $\sqrt{3} \times \sqrt{3}$ R30° state shown in Fig. 2 satisfying the condition, we can locate the inflection point at $x = 1/3$; beyond $x = x/3$ E_g increases inevitably in proportion to nonvanishing $p_1 > N(3x-1)$. The corresponding inflection point in the case of the square lattice is at $x = 1/2$ with the inequality $p_1 \geq \mathrm{Max}[0,\ 2N(2x-1)]$, demonstrating the dependence on the lattice geometry.

In order to show that the inflection point at $x = 1/3$ in the case of the triangular lattice disappears when the nn interaction is not strong enough compared with the competing nnn interaction, we start with the energy expression given by

$$E = V_1p_1 + V_2p_2 \qquad (2)$$

where p_k (k= 1, 2, ···) represents the total number of the k-th neighboring pairs of particles and V's are the corresponding interaction constants. We can prove the following two inequalities [7]:

$$A = 5p_1 + p_2 \geq \mathrm{Max}[0,\ 3N(4x-1),\ 3N(8x-3),\ 3N(12x-6)] \tag{3}$$

$$B = p_1 + p_2 \geq \mathrm{Max}[0,\ N(4x-1),\ N(8x-3),\ N(12x-6)]. \tag{4}$$

We can see that the r.h.s. of the inequalities (3) and (4) has the inflection points at x= 1/4, 1/2 and 3/4. We rewrite the expression (2) in two ways,

$$E = (V_1 - 5V_2)p_1 + V_2 A \qquad \text{and} \tag{5}$$

$$E = (1/4)\ (V_1 - V_2)A + (1/4)\ (5V_2 - V_1)B. \tag{6}$$

By use of the expression (5) and the inequalities (1) and (3) we can prove that if $V_1 > 5V_2 > 0$ $x = 1/4$, the inflection points are 1/3 and 1/2, since the particle configurations satisfying the equality of (1) and (3) at these values of x can be actually found. [7] We can see further by use of the expression (6) and the inequalities (3) and (4) that the candidates of the inflection points in the case $5V_2 > V_1 > V_2$ are x= 1/4 and 1/2; since the configurations satisfying the equality in (5) and (6) can be found at these values of x, we conclude that x= 1/3 ceases to be the inflection point in this range of the interaction constants. We confine ourselves to $x < 1/2$ in the above discussion, since it can be extended easily to $x > 1/2$ by use of the particle-hole symmetry as far as the pairwise interactions are concerned.

The above discussion can be generalized for the energy expression given by

$$E = \textstyle\sum_k V_k p_k + \sum_\ell U_\ell X_\ell + \cdots , \tag{7}$$

where X_ℓ is the total number of appropriately-numbered ℓ-th triplet of particles and U_ℓ is the interaction constant. We assume a finite range of the interactions in the following discussion. Suppose a space spanned by the density x and p's, X's, etc. entering in the energy expression which is of finite dimensions with the above-mentioned assumption. Each inequality restricting the lowest possible value of a linear combination of these variables for a given x defines a plane in this space. The whole set of inequalities will define the polyhedron,whose inside corresponds to allowed values of the variables, p's, X's, etc. On the other hand, the energy expression (7) defines a family of lines on a constant x plane with E as generating parameter. The value of E which produces a contact with the surface of the polyhedron proves to be the ground-state energy,provided that we can find the particle configuration having the values of p's, X's, etc. at the contact point. Note that the surfaces of the polyhedron are defined by the equality in the inequalities.

When we are interested in E_g/N in the limit of infinite N, it is sufficient to find the particle configurations at the vertexes of the polyhedron, since then appropriate two-phase mixtures of them have the p's, X's, etc. corresponding to the edges where the contact between the energy line and the polyhedron is made generally. If we fail to find the configuration at a vertex, on the other hand, we have to replace an inequality at least by finding a new one,which allows us to find the particle configurations at all vertexes of the polyhedron reshaped by it. Thus the key to the success of the analysis is the derivation of appropriate inequalities whose equality is meaningful. In the following,we give a brief summary of the methods of derivation and/or mathematical proofs of the inequalities, referring to [3, 4, 5] for details.

We define a figure to represent a group of s lattice points which are numbered as 1, 2, ···, s. By sliding the figure by unit vectors successively on the lattice, we obtain N figures in total. We use the lattice vector R pointing to the site 1 to denote the figure to which it belongs. The figures can be considered to be snapshots of a given particle configuration on the lattice, when we specify the sites occupied by particles in each figure. To represent the particle configuration on the figure specified by R we introduce the occupation variable $n_i(R)$ with i= 1, 2, ···, s which is equal to 1 if the site i is occupied and to zero otherwise. Note that $n_i(R)$'s are redundant, since a lattice site, and therefore a particle if present, are included in s different figures. We note the following relations:

$$\sum_R 1 = N, \tag{8}$$

$$\sum_R n_i(R) = Nx, \tag{9}$$

$$\sum_R n_i(R) n_j(R) = p_{ij}, \tag{10}$$

where p_{ij} denotes the total number of the particle pairs occupying the sites i and j of a figure; p_{ij} can be expressed in terms of previously-defined p_k when the average over equivalent pairs is taken. We introduce the concept of a function of figure, f(R), by

$$f(R) = \sum_{i<j} \nu_{ij}\, n_i(R) n_j(R) + \sum_i (\mu_i n_i(R) + \lambda_i), \tag{11}$$

where ν_{ij}, μ_i and λ_i are parameters with i and j running from 1 to s. If we can choose the parameters to obtain

$$\sum_R f(R) \geq 0 , \tag{12}$$

we reach an inequality by rewriting the sums of 1, $n_i(R)$ and $n_i(R)n_j(R)$ over R by use of the relations (8), (9) and (10) in terms of N, Nx and p_k's.

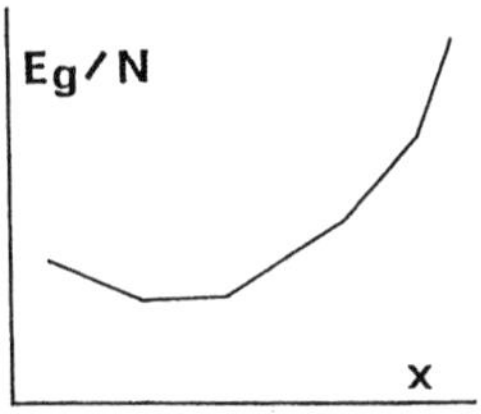

Fig.1. E_g/N vs. x relation of the lattice gas with competing interactions of finite range.

A simple illustrative example of the above argument is given by the choice of f(R) defined by

$$f(R) = (\sum_i n_i(R) - q)(\sum_i n_i(R) - q - 1) \tag{13}$$

which can be reduced to the expression (11) if we note $n_i(R)^2 = n_i(R)$. If the parameter q in (13) is chosen to be an integer, f(R) given by (13) is nonnegative for any particle configuration, since the sum $\sum_i n_i(R)$ is an integer and f(R), the quadratic function of it, can be negative only for non-

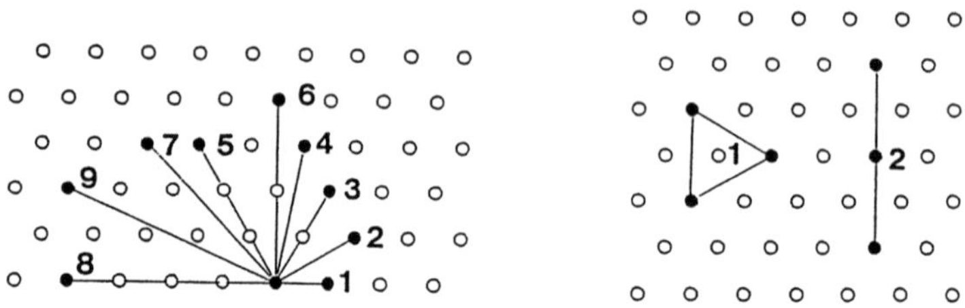

Fig. 2a The definition of k-th neighboring sites with k=1, 2, ···, 9 and examples of triplets, 1 and 2 in the case of the triangular lattice.

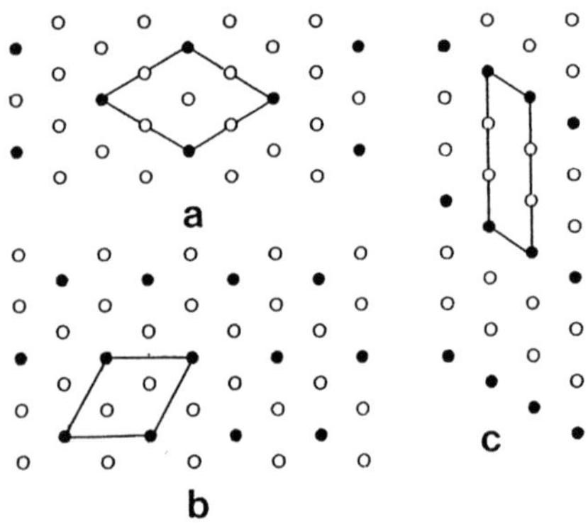

Fig. 2b Examples of ordered states on the triangular lattice. Unit cells are shown. a. (2×2) at $x= 1/4$ with $p_1= p_2= 0$ b. $\sqrt{3} \times \sqrt{3}$ R30° at $x= 1/3$ with $p_1= 0$ and $p_2= N$ c. (1×3) at $x= 1/3$ with $p_1= N/3$ and $p_2= 0$.

integral values of the sum between q and q+1. An extension of this argument is made by another choice of f(R) given by

$$f(R)= (\textstyle\sum_i \nu_i n_i(R) - q_1)(\sum_i \nu_i n_i(R) - q_2) \tag{14}$$

which is non-negative for any particle configuration with integral values of parameters ν_i if integers q_1 and q_2 correspond to two successive integral values of $\sum_i \nu_i n_i(R)$.

The inequalities (3) and (4) are derived by use of the expression (13) and figures shown in Fig. 3; the integral parameter q is chosen so as to result in the maximum of the lower limit for a given value of x. The expression (14) is used in the derivation of the inequalities of linear combinations involving negative coefficients, for example, $-p_1+p_2 > \mathrm{Max}[-Nx, N(x-1)]$ in the triangular lattice case. [7]. The expressions (13) and (14) are non-negative for all figures. We can extend the argument further to the case where f(R) takes negative values for some figures but the sum over R is non-negative. Such an extension, which takes into account the lattice geometry in a wider region than a single figure, is needed in the examples discussed in sections IV and V. We have neglected the more-than-two particle interactions in the above discussion. We can modify the expression (11) to include them easily, however, though details are omitted here. Finally, we remark that the derivation of the inequalities seems to belong to the field of combinatorial geometry, in which the solution has to be worked out in each case.

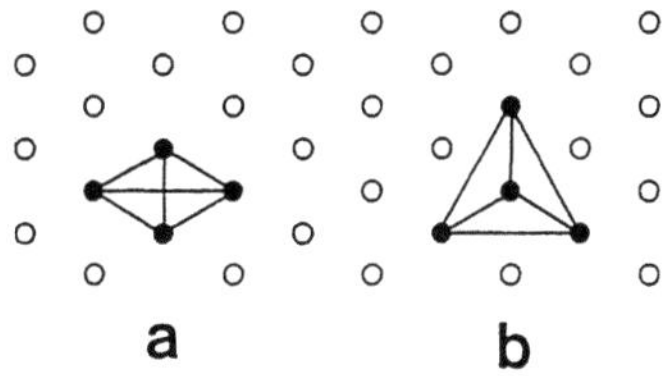

Fig.3. Examples of Figures used in the derivation of inequalities. (a) the figure for the inequality (3). (b) that for (4)

We note that the space spanned by the interaction constants is divided into regions each of which corresponds to a characteristic sequence of ordered states for varying x. Conversely we can get information about the interaction from an observed sequence of ordered states. An example of such a deduction is given by the case where the nn interaction on the triangular lattice is strongly repulsive to forbid the nn pair. By use of the inequalities we can prove that the $\sqrt{3} \times \sqrt{3}$ R30° state is the unique particle configuration realizable at x= 1/3 under the condition $p_1 = 0$. The ordering itself, therefore, does not need an attractive nnn interaction to occur. If we have a repulsive nnn interaction and we can neglect more distant neighbor interactions, however, we should observe the (2×2) state (see Fig. 2) at x= 1/4 which satisfy the conditions $p_1 = 0$ and $p_2 = 0$. If the nnn interaction is attractive, on the other hand, x= 1/3 is the only inflection point.

III. A Summary of the Ground-State Analyses of the Two-Dimensional Lattice Gas Models

We give here a brief summary of the analyses so far carried out before going into more detailed discussions of specific examples in sections IV and V.

A. Square lattice The ground-states are rigorously determined in the whole ranges of interaction constants and density x for up to third neighbor pairwise interactions. [4] With the nn and nnn interactions the simple method of deriving the inequalities by use of (13) and (14) suffices. [2] Inclusion of the third neighbor interaction, however, demands an improved derivation which is capable of taking into account the lattice geometry in a wider region than the previous one. Two new inequalities are added as the result. [4]

B. Triangular lattice The ground-states are rigorously determined in the case of the nn and nnn pairwise interactions. [7] When the third neighbor interaction is added, the analysis fails to determine the ground-states rigorously in the range $1/3 < x < 1/2$, though conjectures are made there. [8]

The analysis of the triangular lattice has been extended recently in two directions with the purpose of discussing the surface ordering. One is to take into account more-than-two particle interactions which destroy the particle-hole symmetry between $x < 1/2$ and $x > 1/2$. Three kinds of the many particle interactions are introduced,in addition to the nn and nnn pairwise interactions. The first one is based on the assumption that the energy of a nnn pair of particles increases when the nearest neighboring sites shared by the pair (see Fig. 4) are occupied by one or two particles. The energy increase is assumed to be independent of the number of particles (1 or 2) at the nn sites. Then the many-particle interaction is expressed as a combination of a three particle interaction and a four particle one. The analysis can be carried out rigorously by deriving the inequalities including the numbers of the triplets and the quartets. It is concluded that besides

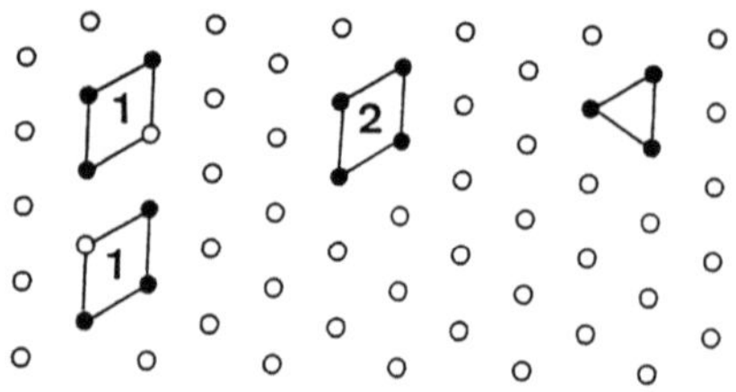

Fig. 4. Definition of three kinds of the many particle interactions (see text). a. The particle configurations 1 and 2 of the rhomb are assumed to have a common extra energy besides the pair energies. b. The extra energy of the configuration 2 is twice that of the configuration 1. c. The triplet occupying the mutually nearest neighboring sites has an extra energy.

the removal of the particle-hole symmetry, the many-body interaction gives rise to new ordered states if it is strong enough, compared with the pair interactions. In the second model we drop the four particle interaction from the energy of the first model; the energy-increase of a nnn pair of particles is then proportional to the number of particles at the nn sites shared by the pair. In the third model we take account of the three particle interaction among the mutually nearest neighboring sites. The analysis of these two models can be carried out rigorously also. The three particle interactions can give rise to new ordered states. Details will be reported in our future publication.

Another extension of the analysis of the triangular lattice is made in the case where a sufficiently strong repulsive nn interaction results in $p_1 = 0$. The purpose of the discussion is to establish the lattice gas model of the (7×7) and related reconstruction of the (111) surface of Si, Ge and Ge-Sn alloys. [6] We discuss this subject in section V. The ground states for up to fourth neighbor interactions under the condition $p_1 = 0$ are determined rigorously for $x < 1/4$.

C. Honeycomb lattice As an example of non-Bravais lattices this case was discussed by Kudo and Katsura. [9] However, they failed to determine the ground states in the case $V_1 > 2V_2 > 0$ for $x > 1/4$. As will be discussed in the next section, three kinds of infinite series of the ground states are discovered by the present author in this case. The analysis is now complete for the case where up to third neighbor interactions are taken into account. [4]

D. Kagomé lattice This is another example of the non-Bravais lattices which has been analyzed recently by Okamoto and the present author. We shall publish the result in a future publication. No infinite series of the ground states is found in the case where up to third neighbor interactions are taken into account. We distinguish two kinds of the third neighbors: one is the pair not intervened by another site and the other is that intervened by a site at the midpoint.

E. The face-centered rectangular lattice (the (110) surface of bcc crystal)
This model is the first one to which the method of geometrical inequalities was applied. The method was originally used to determine the magnetization process of the antiferromagnetic Ising model in which intermediate phases may appear between the initial antiferromagnetic state in small external field and the final ferromagnetic state in strong field. [2] The model was adopted to represent a crystal of $CoCl_2 \cdot 2H_2O$ in which such a magnetization process was found for the first time by KOBAYASHI and HASEDA. [10] In this case an Ising spin corresponds to a linear chain of Co atoms

on which spins are coupled ferromagnetically. The model was later analyzed to discuss the ordering of Na atoms on the (110) surface of W which was found experimentally at x= 1/4, 1/6 and 1/8. The ground states are determined with up to fourth neighbor pair interactions in the whole range of x. In the range of $1/4 > x$ up to seventh neighbor interactions are taken into account in the analysis. [11]

IV. The Honeycomb Lattice

We mention briefly the difficulty encountered in the analysis based on a simple derivation of the inequalities. [5] We can prove easily the following inequalities for $x < 1/2$:

$$2p_1 + p_2 \geq \text{Max}[0, N(4x-1)], \tag{15}$$

$$4p_1 + p_2 \geq \text{Max}[0, N(4x-1), N(9x-3)], \tag{16}$$

$$p_1 \geq 0 , \tag{17}$$

$$p_2 \geq \text{Max}[0, N(3x-1)]. \tag{18}$$

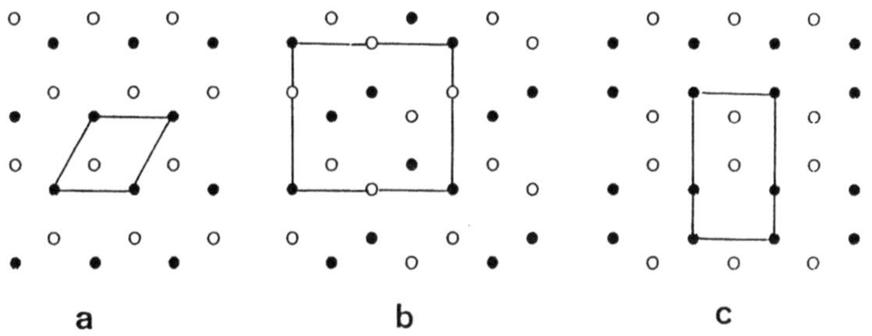

Fig. 5. The ground states of the honeycomb lattice at x= 1/2. a. S_1 b. S_2 c. S_3.

We consider the nn and nnn pair interaction in the following. The third neighbor interaction, however, is important to remove the degeneracy among the particle configurations of lowest energy, though it is not essential for the occurrence of the infinite series. The inequalities given above determine straightforwardly the ground states at x= 1/2 to conclude that the state S_1 shown in Fig. 5 is of lowest energy for $V_1 > 4V_2$ with $p_1 = 0$ and $p_2 = 3N/2$ satisfying the equality in (16) and (17) and that S_2 (for $V_3 > 0$) or S_3 (for $V_3 < 0$) shown in Fig. 5 is of lowest energy for $4V_2 > V_1 > 0$ with $p_1 = N/4$ and $p_2 = N/2$ satisfying the equality in (15), (16) and (18). The inequality (16) rather than (15) is essential to determine these conditions on the interaction constants. We can see easily that if $p_1 = 0$ (16) becomes more restrictive for p_2 for $x > 2/5$ than (15). When we construct the polyhedron discussed in section II we encounter the difficulty at the vertex given by x= 2/5, $p_1 = 0$ and $p_2 = 3N/5$ that we cannot find the corresponding particle configuration. As a result we cannot determine the gound states with the inequalities shown above in the region $V_1 > 2V_2 > 0$ and $1/2 > x > 1/3$ which becomes $V_1 > 2V_2 > 2V_3 > 0$ or $V_1 > 2V_2 - 2V_3$, $2V_2 > V_3$ and $V_3 < 0$ shown in Fig. 6.

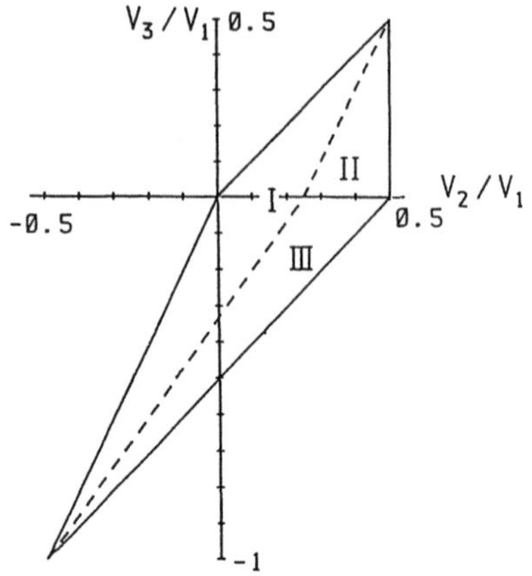

Fig. 6. The interaction constant space of the honeycomb lattice showing the regions under discussion. The states S_1, S_2 and S_3 shown in Fig. 5 are the ground states in the regions I, II and III, respectively. The infinite series of the ground states appear in the region I, while they are terminated at a finite number in the regions II and III.

The two triangular sublattices which make the honeycomb lattice will be labelled with A and B, respectively. Imagine the situation $V_1 \gg V_2$ which imposes the condition $p_1 = 0$. In the ground state S_1 at $x=1/2$ shown in Fig. 5 one of the triangular sublattices is fully occupied by particles. When the density x is smaller than 1/2, we can divide the lattice into two regions, type A and type B, in such a way that in a given region particles are on the same sublattice and connected at least by a nnn line. Note that with the condition $p_1 = 0$ the regions A and B do not penetrate into each other. The regions A and B are separated by a zone called dislocation, which is bordered by two parallel lines connecting next nearest neighboring sites on the sublattices A and B. Particles on the border of the A region are generally third neighbors of those on the border of the B region. It can be proved that a wider separating zone leads us to a state of higher energy. See Fig. 7 which illustrates the situation. Because of the lattice geometry the

Fig. 7. Dislocations and their junctions. The hatched regions are the dislocations or walls separating the regions where one of the triangular sublattices, A or B is occupied by particles (see text). Junctions labeled with 1, 2 and 3 appear in the ground states.

dislocations cannot run parallel but always make an angle of 60° with each other. Thus the shape of a region either A or B is an equilateral triangle, except for the fact that it can be connected to another triangle of the same label at the vertex. Around each vertex of the triangles, dislocations meet with each other to make a junction. Figure 7 shows some examples of the junctions, of which the types 1 and 2 are energetically favored in the case of $V_1 \gg V_2$. When V_1 becomes somewhat smaller, the type 3 junction can appear in the ground states.

Three kinds of infinite series, T, H and R of the states shown in Fig. 8 which make use of the junctions 1, 2 and 3, respectively are proved to be the ground states in the range $1/3 < x < 1/2$ (the H series starts at $x= 1/4$) after a tedious analysis. [5] They are the ground states at the inflection points of the E_g/N vs. x relation in certain regions of the space spanned by V_1, V_2 and V_3. We conclude that the series of the ground states is infinite in the region defined by $V_1 - 4V_2 + 2V_3 > 0$ and $V_2 > V_3$ with $V_3 > 0$ and also in that given by $V_1 - 4V_2 + 3V_3 > 0$ and $2V_2 > V_3$ with $V_3 < 0$. [5] In the rest of the region shown in Fig. 6 where the state S_2 or S_3 is of lowest energy at x= 1/2 the series terminates at a finite number. Note that T_n, H_n and R_n with infinite n correspond to S_1. We refer to [5] for further details, including mathematical proofs which involve the derivation of two sets of an infinite number of new inequalities.

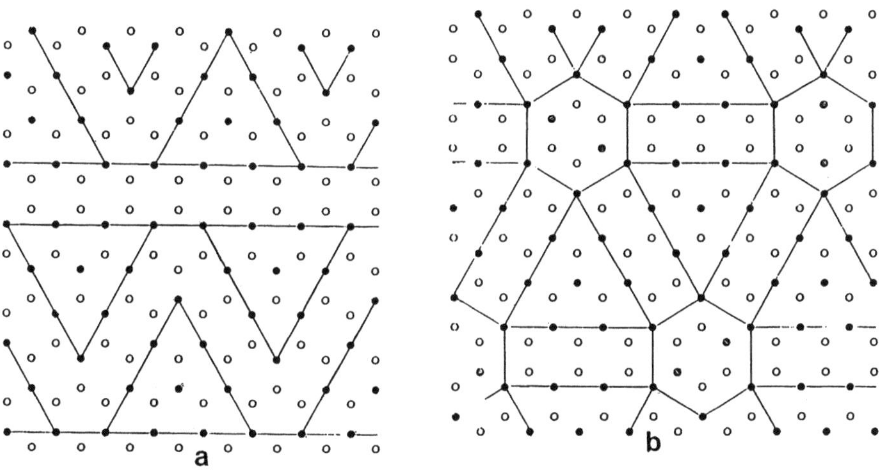

Fig. 8. The three series of the ground states of the honeycomb lattice. The number of lattice sites on a side of an equilateral triangle, n defines the n-th state of each series. a. An example of the T series states. b. Examples of the H and R series states. The hatched circles in junctions of dislocations are open circles (vacant sites) in the H series, and solid circles (sites occupied by particles) in the R series.

The fcc lattice is a stacking of three kinds of the triangular nets specified by A, B and C hereafter. Suppose the (111) surface of a crystal is an A plane. When adsorbates take indiscriminatively the B or C sites on the surface, the lattice for them is given by a honeycomb lattice. Another possibility of realizing the honeycomb lattice is given in the case where two kinds of adsorbates are present on the triangular lattice. When one kind of adsorbates make a $\sqrt{3} \times \sqrt{3}$ R30° ordered state, the rest of the sites available for another kind of adsorbates will make a honeycomb lattice. We notice here that if the infinite series is observed in a real system, it may appear as a stepwise growth of islands of the shape of an eqilateral triangle,which correspond to the state S_1. Then the configuration of the islands will distinguish between the T and H series. In the case of the R series the inherent disorder within the type 3 junction discussed in [5] will complicate the interpretation.

V. The (7×7) and Related Reconstructions of the (111) Surface

The stable structure of the (111) surface of Si at low temperatures (<1000K) is the well-known (7×7) structure. [12] The corresponding reconstruction of the (111) surface of Ge is the (2×8) structure below 500K. [13] The (2×8), (7×7) and (5×5) phases are found successively in Ge-Sn alloys when Sn concentration is increased. [14] Recently Binnig, Rohber, Gerber and Weibel have determined the (7×7) structure shown in Fig. 9 by the scanning tunneling microscopy method [12]; the solid circles in Fig. 9 represent the spots where the surface plane is raised. They have proposed the 'adatom' model which assumes the presence of an additional atom there on the surface layer. Though the local structure around the 'adatom' seems to be still controversial [15,16], we may represent it as a particle of the lattice gas to discuss the overall structure of the phase. From this point of view we can show that the (2×8) and (5×5) reconstructions which were studied previously [13,14] are the ground states of the lattice gas model with the same interaction constants as those which stabilize the (7×7) one. After establishing the lattice gas model on the basis of the ground state analysis we extend the discussion to finite temperatures to discuss, for example, the phase-transition and short range orders in the disordered state above the transition temperature. Since this research has not been completed yet and since we have reported preliminary results elsewhere [6,17], we give below a brief summary of the discussion.

From Fig. 9 we see that the (2×8) structure can be regarded as a uniaxially modulated state of the (2×2) structure, while the (7×7) and (5×5) ones are biaxially modulated states of the same structure. The latter states belong to the series of the (n×n) states with n=3.5.7.9, ... which ends at the (2×2) state with x= 1/4 in the limit of infinite n. When the particle arrangement in the modulation zone is extended to the whole lattice, we reach the (2×4) state shown in Fig. 9. Thus we can understand the condition for the appearance of the (2×8) structure concluded from the ground-state analysis that 1) the (2×2) state is the ground-state with up to fourth neighbor interactions, 2) it is almost degenerate with the (2×4) one and 3) a repulsive sixth neighbor interaction and/or an attractive seventh neighbor interaction can stabilize the (2×8) one. We can show further that up to eleventh neighbor interactions will not lead us to modulated structures other than the (2×8) one. In order to stabilize the biaxial modulation in the (7×7) and (5×5) states, we need additional conditions to those for the (2×8) state, because other types of modulations of the (2×2) state are possible for $x < 1/4$. The additional condition can be expressed as that for stabilizing the (3×3) state shown in Fig. 9 which is the first of the series of the (n×n) mentioned above. A competing modulation is mentioned in [17]. One of the conclusions drawn from the lattice gas model with up to seventh neighbor interactions is that the (n×n) states with n larger than 7 are excluded from the ground states when the modulation is stabilized. This is because the states with $n > 7$ are too close to the (2×2) state with less density of modulation, which does not favor them energetically compared with two-phase mixtures between the (7×7) and (2×8). The reasoning mentioned above is substantiated by the analysis based on the inequalities,whose derivation is quite complicated. We have reported the result of the analysis in [6]; some supplementary discussions and corrections of errors in [6] are given in [17].

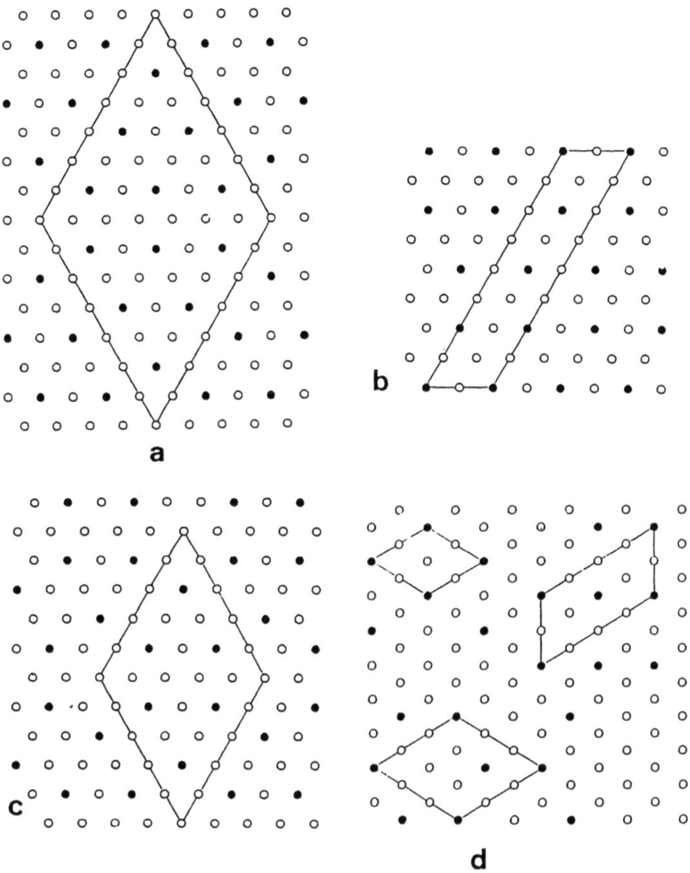

Fig. 9. The surface reconstructions of the (111) surface of Si, Ge and their alloys and related states. a. (7×7) b. (2×8) c. (5×5) d. (2×2) at x= 1/4 (upper left), (2×4) at x= 1/4 (upper right) and (3×3) at x= 2/9.

A Monte Carlo computer simulation based on the lattice gas model is being carried out at present with the purpose of discussing in the first place the disordered state above the transition-temperature of the (2×8), (7×7) and (5×5) states. Preliminary results of this research are briefly mentioned in [17]. We shall report in a future publication more details of the research on the ground states as well as the disordered one.

VI. Concluding Remarks

We have discussed the application of the lattice gas model to complicated lattice orderings which need a careful analysis of the ground states. Extensions of the discussion in various directions are conceivable. The possibility of more-than-one kinds of adsorption sites is an example. The problem of more-than-one kinds of adsorbates is another example. The latter is actually being investigated by Tonegawa. [18] Finally we mention that applications of the present ground-state analysis of the lattice gas model were made so far mostly to the three-dimensional cases of ordered alloys and compounds. [19-21]

Acknowledgements

The author would like to express his sincere thanks to Mr. M. Okamoto for valuable discussions on the (111) surface reconstruction of Si, Ge, etc. Thanks are also due to Drs. T. Tonegawa, M. Kaburagi, J. Igarashi, S. Ino and A. Yoshimori for valuable and enlightening discussions. He thanks Mrs. A. Egusa for her efforts in typing the manuscript. The present research has been supported partially by a Grant-in-Aid for Scientific Research from the Ministry of Education, Science and Culture.

References

1. W. Selke, K. Binder and W. Kinzel, Surface Sci. 125, 74 (1983)
2. J. Kanamori, Prog. Theoret. Phys. 35, 16 (1966)
3. M. Kaburagi and J. Kanamori, Prog. Theoret. Phys. 54, 20 (1975)
4. J. Kanamori and M. Kaburagi, J. Phy. Soc. Jpn. 52, 4184 (1983)
5. J. Kanamori, J. Phys. Soc. Jpn. 53, 250 (1984)
6. J. Kanamori, Solid State Commun. 50, 363 (1984)
7. M. Kaburagi and J. Kanamori, Jpn. J. Appl. Phys. 13 Suppl. 2 part 2 (Proc. 2nd Intl. Conf. Solid Surfaces), 145 (1974)
8. M. Kaburagi and J. Kanamori, J. Phys. Soc. Jpn. 40, 291 (1976)
9. T. Kudo and S. Katsura, Prog. Theoret. Phys. 56, 435 (1976)
10. H. Kobayashi and T. Haseda, J. Phys. Soc. Jpn. 19, 765 (1964)
11. M. Kaburagi and J. Kanamori, J. Phys. Soc. Jpn. 43, 1686 (1977); M. Kaburagi, J. Phys. Soc. Jpn. 44, 394 (1978)
12. G. Binnig, H. Rohrer, Ch. Gerber and E. Weibel, Phys. Rev. Letters 50, 120 (1983)
13. P. W. Palmberg and W. T. Peria, Surface Sci. 6, 57 (1967)
14. T. Ichikawa and S. Ino, Surface Sci. 105, 395 (1981)
15. L. C. Snyder, Z. Wassermann and J. M. Moskowitz, J. Vac. Sci. Tech. 16, 1266 (1976)
16. M. Aono, R. Souda, C. Oshima and Y. Ishizawa, Phys. Rev. Letters 51, 801 (1983)
17. J. Kanamori, J. de Phys. (Proc. Colloq. A. Blandin) to be published
18. T. Tonegawa, Private communication
19. J. Kanamori and Y. Kakehashi, J. de Phys. 38, C7-274 (1977); J. Kanamori, AIP Conf. Proc. 53, 117 (Intl. Conf. Modulated Str.) (1979)
20. M. Kaburagi, AIP Conf. Proc. 53, 127 (1979)
21. A. Finel and F. Ducastelle, Proc. of 2nd Intl. Conf. on Phase Transformation in Solids, June 27 - July 1, 1983 Crete (to be published)

Electronic Structure of Metal Overlayers on Si(111) Surface

H. Nagayoshi

Department of Material Physics, Faculty of Engineering Science
Osaka University, Toyonaka, Osaka 560, Japan

1. Introduction

There has been a long history of research into the electronic properties of metal-semiconductor interfaces. Recently a large amount has dealt with semiconductor surfaces with chemisorbed metal atoms in the sub-monolayer range. For such systems various types of superlattice structures are observed. Many experimental works are devoted to the determination of atomic arrangement [1-4] in these systems by the use of such methods as low energy electron diffraction (LEED), ion scattering spectroscopy (ISS), extended X ray absorption fine structure (EXAFS), and photoelectron diffraction. Their electronic stucture [5-7] is also extensively investigated by, for example, angle-resolved ultraviolet photoemission spectroscopy (ARUPS).

At this stage it is of great interest whether the proposed geometry in the former analyses is consistent with the results of the latter in the theoretical point of view. To some extent discussions can be made on the basis of chemical bond picture. Quantitative and conclusive discussions, however, require first principle calculations of the one-electron structure by the use of each geometry proposed. Such calculations will also predict a lot of unknown properties of these systems, and provide convenient models on which advanced theoretical analyses are to be made.

In this work we present a study based on numerical calculations with the ion-core pseudopotential and the local density functional approximation. This approach to the problem has been useful in a wide variety of semiconductor surfaces with reconstruction and chemisorbed atoms [8-12]. Some numerical results are given here for the $\sqrt{3}\times\sqrt{3}$ superlattice structures observed in the Al and Ag chemisorption on the Si(111) surface. The former system [6,11] is an example for which theoretical analysis can be made in an established routine, and for the latter there are a good deal of experimental data [1-7] available at present.

2. Si(111):Al $\sqrt{3}\times\sqrt{3}$ by Semiempirical Pseudopotential Method

Only a few experimental data are available for this surface. HANSSON et al [6] obtained the dispersion of surface states in the surface Brillouin zone (SBZ) from their ARUPS experiment. They also proposed that the Al atoms are located at the threefold hollow sites, with the fourth layer Si atoms beneath, on the basis of the azimuthal angle-dependence of the emission pattern. There has been no such direct structure analysis as EXAFS for this system.

We present here self-consistent pseudopotential calculations of this system in a similar way and using the same ionic potential as ZHANG and

SCHLUTER used in their work [12] on the 1×1 Al overlayers. Pseudopotentials of Si^{4+} and Al^{3+} ions, which are the input of the calculation, are expressed by the following 4 parameter form

$$v(q) = \frac{a_1}{q^2}(a_2 + a_3q^2)\exp(a_4q^4) \qquad (1)$$

in the q space. The parameter values are determined by fitting the ionic energy levels to spectroscopic data. In this sense, this potential is semi-empirical. This type of (local) pseudopotential has been successful in various bulk and surface problems [8-12].

Self-consistency is achieved simply in the Slater's Xα scheme with α=0.79 for the exchange-correlation term. Despite many discussions on the density functional form, we believe that the essential feature of the electronic structure does not depend on the choice of the functional form, unless we are concerned with a very delicate quantity such as the total energy of the system.

Among the various adsorption sites of Al atoms for which we made calculations, we present here the results for the threefold sites by HANSSON et al with Al atom at height z_{Al}=1.47, 2.57, 3.64, and 4.88a_B. As usual, calculations are performed for the slab of Si with Al overlayers on both sides. In the plane wave representation, the potential is truncated at $q=2.8a_B^{-1}$, and the wave functions at $q=2.3a_B^{-1}$. For the latter, plane waves with $q>1.5a_B^{-1}$ are included in the perturbational way. In Fig. 1 we show the E-k relation for $z_{Al}=3.64a_B$ in the energy region of valence band maximum. Here our slab contains 14 Si layers. Essential feature is the same for 20 Si layers, for which the bands of surface states have smaller energy-splitting (<0.03eV) due to the overlap through the slab.

Three bands of surface states are found in the gap of substrate. We find that they have a close resemblance to those of the ideal clean surface. The presence of Al pushes two of them downward by ∿0.6eV and the remaining one a little upward as a whole; characteristic shape of the bands is left unchanged. For Al height $z_{Al}>3.64a_B$, decreasing z_{Al} causes considerable downward shift of the lower surface states (LSS), and for $z_{Al}<3.64a_B$ upward shift and increasing dispersion of the upper surface states (USS).

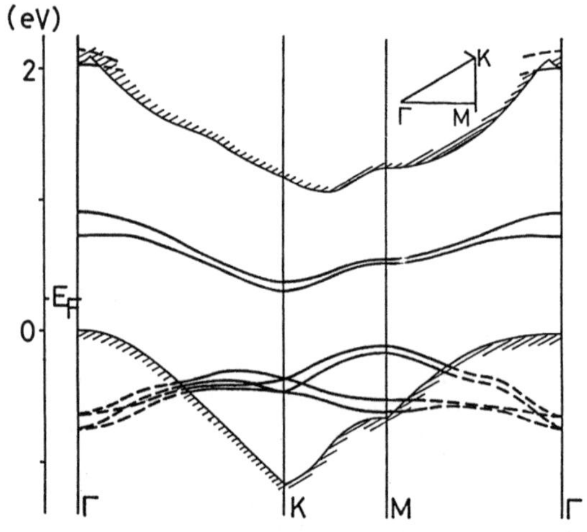

Fig.1 Energy-dispersion of surface states of Al:Si system ($z_{Al}=3.64a_B$)

Very recently NORTHRUP [11] presented a similar calculation using the so-called norm-conserving pseudopotential for the input potential, with the total energy calculation of this system. His results for the energy band diagram are essentially the same as ours. He claimed that the splitting of the LSS and USS is due to the favorable and unfavorable coupling of the top layer Si dangling bond, respectively, with the p_x and p_y orbitals of the adsorbed Al atom. We think, however, that this explanation is not suitable. In Fig. 2 we present the contour map of wave functions of the LSS and USS at the K point. The LSS is, as shown, essentially a pure Si dangling bond state, deformed very little by the mixing of Al p_x and p_y components. On the other hand, the USS contains large Al components. We are going to consider the character of the LSS and USS in detail here, on the basis of our numerical calculations.

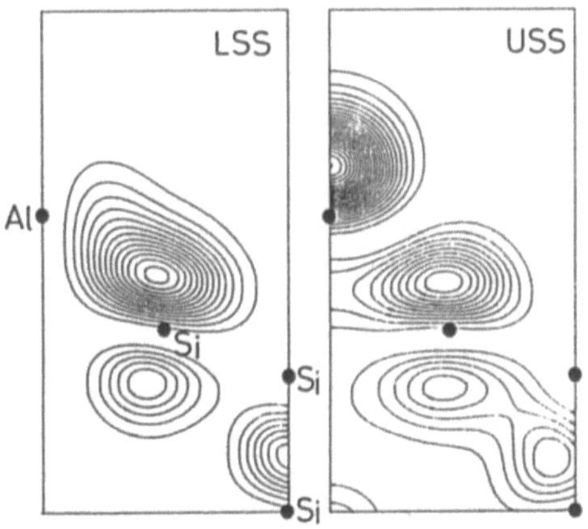

Fig. 2 Contour map of $|\varphi(r)|^2$ of the LSS and USS

First we consider the splitting of the LSS and USS at K. As mentioned, the LSS is essentially made up of the Si dangling bond orbitals. Their combination takes the form $\varphi_1+\omega\varphi_2+\omega^2\varphi_3$ for the dangling bonds $\varphi_1 \sim \varphi_3$ around each Al atom, where ω denotes the imaginary cube root of unity. For this combination, the Al-induced potential $v(r)$ leads to the energy shift v_0-v_1, where

$$v_0 = \int \varphi_i(r)v(r)\varphi_i(r)d^3r, \quad v_1 = \int \varphi_i(r)v(r)\varphi_j(r)d^3r \quad (i\neq j). \qquad (2)$$

Similar consideration for the USS gives the shift v_0+2v_1. Thus the splitting of the LSS and USS is essentially caused by the three-center integral v_1, whose sign is expected to be positive.

For the USS, mixing of Al orbitals is to be added to the above discussion. We prepare the Al s and p_z orbitals and Si p_z (dangling bond) orbital. In Fig.3(a-d) we show the essence of our interpretation. The orbitals (a) and (b) with bonding coupling of the Si p_z and Al s-p_z mixed orbitals are stabilized a good deal, and are expected to exist deep in the valence band. The orbital (c) is expected to have very high energy due to the strong anti-bonding character, and (d) with weak anti-bond coupling corresponds to the USS in the gap. The mixing ratio of Al s and p_z orbitals depends greatly on z_{Al} in the range $z_{Al}<3.64a_B$, which gives rise to the strong dependence of the USS on z_{Al}.

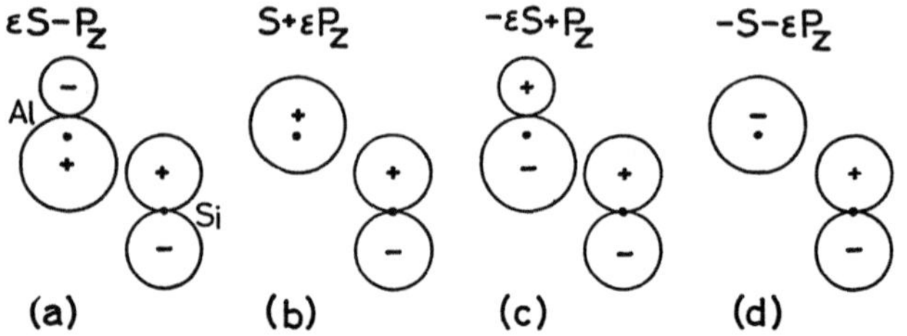

Fig. 3 Schematic representation of the four Al-induced states

The above interpretation is also consistent with the electron filling. Each Al atom supplies two electrons to the bonding states (a) and (b), and one to the dangling bond state LSS, leading to the fully occupied LSS band.

Our results for the LSS are regarded to be consistent with HASSSON et al's UPS data [6]. The surface states they obtained lie in energy between the dangling bond and back bond states of the ideal surface, and the shape of dispersion is similar to the former in the zone corner of the 1×1 SBZ. These are in agreement with our LSS. However, to locate the LSS to be 1.5eV lower than E_F as they reported, we should put E_F above the USS for $z_{Al}=3.64a_B$ or within it for $z_{Al}=2.57a_B$. Special consideration will be necessary for the discussion on the position of E_F measured in UPS. Self-interaction correction [13] to the density functional formalism or the many-body effect [14] might explain the apparent downward shift of the occupied surface states with respect to E_F.

3. Si(111):Ag $\sqrt{3}\times\sqrt{3}$ by the Mixed Basis Method with Norm-Conserving Pseudopotential

Compared with the Al:Si system, we have much experimental information on the Ag adsorbed surface. Recent structural analyses by ISS [1], LEED [2], EXAFS [3], and X ray photoelectron diffraction [4] claim that the Ag atoms are buried underneath the top layer of Si. There are also many ARUPS [6,7] data available. Here we would like to study this system theoretically and compare it with these experimental results.

For this system the Ag 4d electrons play an essential role, and we need to introduce the ℓ-dependence of pseudopotential. For this purpose the norm-conserving pseudopotential by BACHELET et al [15] is useful. It is of the form

$$v(r) = -\frac{2Z}{r}\sum_i c_i \,\mathrm{erf}(\sqrt{\alpha_i}\, r) + \sum_\ell |\ell\rangle \sum_i (A_i^{(\ell)} + A_{i+3}^{(\ell)} r^2)\exp(-\alpha_i^{(\ell)} r^2)\langle\ell| \quad (3)$$

with parameter values c_i, α_i, $A_i^{(\ell)}$, and $\alpha_i^{(\ell)}$ given in their table. Spin-orbit interaction is completely neglected here for simplicity. To express the wave function we use the so-called mixed basis set; namely, we prepare plane waves as well as five d-like atomic functions φ_i per each Ag atom, which are chosen to be the solution of a free atom. For mathematical convenience we modify their tail so that it goes smoothly to zero at the atomic radius $R_c=2.52a_B$ [16]. Since the atomic functions centered at different sites do not overlap, many-centered integrals do not appear. All the overlap integrals are evaluated by simple integration such as

$$\int_0^{R_c} \varphi(r)^2 v_a(r) r^2 dr, \quad (4)$$

which describes the energy position of Ag 4d-like states relative to the substrate valence band. Here $v_a(r)$ in (4) denotes the atomic part of the self-consistent potential. Self-consistency is achieved by PERDEW and ZUNGER's expression [13] for the exchange-correlation potential. Also, in the calculation of valence electron density and screening potential, we separate the atomic part which goes to zero at R_c around the Ag atom from the remaining plane wave part expressing the smooth variation of the quantity considered. The details of the scheme will be presented elsewhere.

In actual calculations the cut-off of plane-wave expansion is taken as $5.0a_B^{-1}$ for the core potential and $3.1a_B^{-1}$ for the screening potential. In the wave-function expansion cut-off wavenumber $2.5a_B^{-1}$ is used. In the ℓ expansion of the atomic screening potential, we keep only the $\ell=0$ component here.

In the following, we present preliminary reports on numerical calculations recently performed. Figure 4 shows one model for which calculations have been made. In this model, layer separations associated with the Ag and first three S_i layers are obtained from the LEED analysis by TERADA et al [2], and the atoms below the third Si layer are located for simplicity at the ideal positions. The lateral positions of Si atoms in the first and second layers are assumed, so that the distance between Si atoms in adjacent layers is equal to the Si-Si bond length in the bulk. Due to the large reduction of second and third Si layer separation, the bond between them is inclined very much from the ideal situation. The slab contains 14 Si layers and two Ag layers.

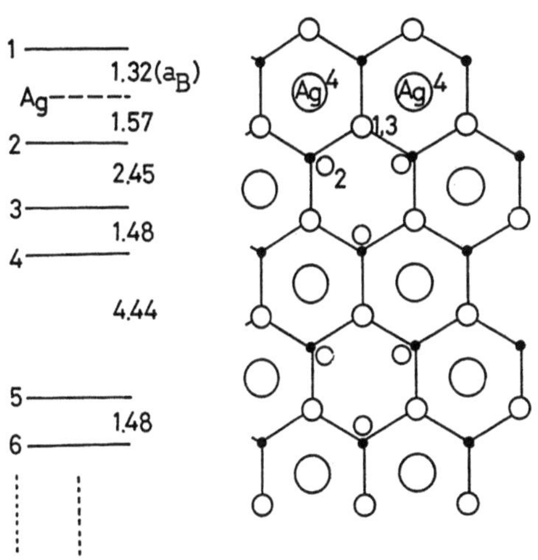

Fig. 4 Model for Ag:Si system used in the calculation

According to the calculated results, there is a group of dense energy levels which have large Ag 4d components in the energy region E_F-5eV ∿ E_F-6eV. This is consistent with the UPS data [6,7]. Since we have not calculated the density of states yet, it is not clear whether this 4d contribution takes a two-peak structure as obtained in UPS and in the DVXα cluster calculation by HOSHINO [17]. Except for these Ag 4d-like levels, the level distribution throughout the valence band seems to exhibit the feature of the valence band of the substrate as a whole. In addition, we find 2∿3 clear surface-state split-off from the bottom of the valence band.

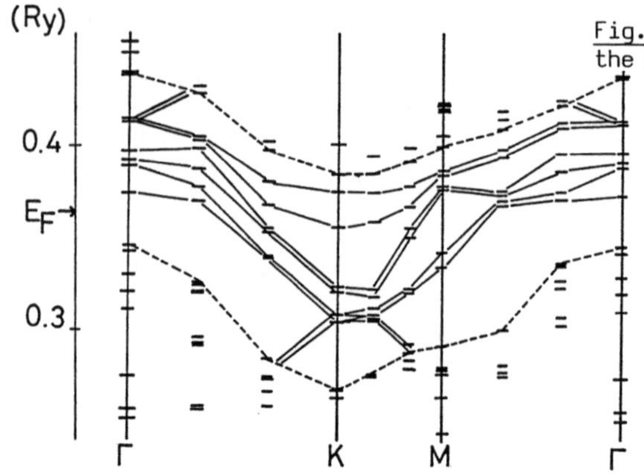

Fig. 5 Energy bands obtained for the model in Fig. 4

The energy band structure near E_F is shown in Fig. 5. The levels in this region do not always take a clear pair-band structure which is characteristic of the surface states in the energy gap of the substrate. As a result there is some ambiguity as to how to connect the obtained levels at each k. It may be that we have to use a thicker slab, which seems to be necessary also for the identification of many resonances between E_F and the Ag 4d peak found in the UPS results [7]. We have not made detailed analyses as to the consistency of our results with ARUPS spectra. A contour map of the states near E_F shows, in general, a large p_z component of the third layer Si atom. The wave function is extended in the direction of the Si-Si bond between the second and third layers in the ideal surface, not in the assumed bond direction in the present model. This suggests that the geomtry assumed here is energetically unfavorable. We are now going on with the calculation for other geometries proposed by EXAFS and photoelectron diffraction.

4. Summary

It has been demonstrated that the established scheme using slab geometry and suitable ionic pseudopotential with self-consistent band calculation is a useful tool for the theoretical investigation of the electronic structure of metal overlayers on semiconductor surfaces. As an example with no localized d electrons, a study of the Al $\sqrt{3}\times\sqrt{3}$ overlayer on Si(111) surface has been presented on the basis of the calculations performed by the ordinary plane wave expansion scheme for wave functions. The results can be reasonably interpreted in the chemical bond picture, and are also consistent with UPS experiments.

For the systems with d electrons, the mixed basis scheme is to be employed throughout the calculation. Preliminary results have been presented for the Si(111):Ag $\sqrt{3}\times\sqrt{3}$, for which many experiments have been done. Although detailed analyses of the obtained results have not been completed yet, it is hopeful that reasonable explanations of experimental data such as the UPS spectra are given in this direction of theoretical approach.

References

1 M. Saitoh, F. Shoji, K. Oura, and T. Hanawa, Japan. J. Appl. Phys. 19 (1980) L421.
2 Y. Terada, T. Yoshizuka, K. Oura, and T. Hanawa, Surface Sci. 114 (1982) 65.
3 J. Stohr, R. Jaeger, G. Rossi, T. Kendelewicz, and I. Lindau, Surface Sci. 134 (1983) 813.
4 S. Kono, H. Sakurai, K. Higashiyama, and T. Sagawa, Surface Sci. 130 (1983) L299.
5 F. Wehking, H. Beckermann, and R. Niedermayer, Surface Sci. 71 (1978) 364.
6 G.V. Hansson, R.Z. Bachrach, R.S. Bauer, and P. Chiaradia, Phys. Rev. Letters, 46 (1981) 1033.
7 T. Yokotsuka, S. Kono, S. Suzuki, and T. Sagawa, Surface Sci. 127 (1983) 35.
8 M. Schluter, J.R. Chelikowsky, S.G. Louie, and M.L. Cohen, Phys. Rev. B12 (1975) 4200.
9 K.M. Ho, M.L. Cohen, and M. Schluter, Phys. Rev, B15 (1977) 3888.
10 H. Nagayoshi and M. Tsukada, Surface Sci. 116 (1982) 163.
11 J.E. Northrup, Phys. Rev. Letters 53 (1984) 683.
12 H.I. Zhang and M. Schluter, Phys. Rev. B18 (1978) 1923.
13 J.P. Perdew and A. Zunger, Phys. Rev. B23 (1981) 5048.
14 E. Louis, F. Flores, F. Guinea, and C. Tejedor, Solid State Commun. 44 (1982) 1633.
15 G.B. Bachelet, D.R. Hamann, and M. Schluter, Phys. Rev. B26 (1982) 4199.
16 L. Kleinman and Ed Caruthers, Phys. Rev. B10 (1974) 3213.
17 T. Hoshino, Surface Sci. 121 (1982) 1.

Commensurate Structures of Ionic Adlayer with Repulsive Long Range Interaction

Kazumasa Shinjo and Tomohei Sasada*

Institute for Solid-State Physics, University of Tokyo
7-22-1 Roppongi, Minato-ku, Tokyo 106, Japan
and
*Sagami Institute of Technology, Fujisawa 251, Japan

A study is made of the ground state of an ionic adlayer with repulsive long range interaction. The method for constructing commensurate structures is presented for the simplest cases of coverage. The lower bound for the ground-state energy is given as a function of the coverage. A misfit between an ordered structure due to the intrinsic interaction between adatoms and the substrate structure is discussed. It is found that for the square lattice of the substrate,the adlayer is frustrated over a whole range of coverage, and for the regular triangular lattice,the adlayer is unfrustrated for a certain class of coverage.

1. Introduction

Structures and phase-transitions[1] of adlayers on metallic surfaces have continued to be the subject of extensive studies during the last two decades. Among the most interesting adlayers are those with a largely polar bonding of adatoms with metallic substrates, e.g., adlayers of alkali and alkaline earth elements, and of halide elements on various metals.

When atoms are adsorbed on metallic surfaces, they experience interaction not only with the substrate, but also a lateral one between adatoms. In the case of ionic adsorption, the adatom is in a charged state, and forms, together with its image,a dipole perpendicular to the surface. The lateral interaction between adatoms is a dipole-dipole repulsive force for small degrees of coverage. This type of interaction is simple and definite,since it does not depend on details of electronic states of adatoms. In spite of the simplicity of the interaction, only a few theoretical studies have been made for understanding structural consequences of lateral interaction in ionic adlayers.

A decade ago, KABURAGI and KANAMORI [2] investigated commensurate structures on the basis of the lattice gas model where the interaction is assumed to be of an extended but finite range. Subsequently, POKROVSKY and UIMIN [3] took into account an infinite range of interaction,to show that the dipole-dipole interaction is responsible for the formation of commensurate structures with a very large period,which are observed at very low coverages. Their method for constructing commensurate structures of adatoms is a direct minimization of the dipole-dipole interaction energy for all possible configurations of adlayers.

In this paper [4] we are concerned with the ground states of ionic adlayers,to present a different method for constructing commensurate structures,and to study structural and energetical properties of this system. We proceed now to describe the ionic adlayer in terms of a lattice gas model with a long-range repulsion. The Hamiltonian is

$$H = \sum \mu n_i + \frac{1}{2}\sum{}' J_{ij} n_i n_j, \tag{1}$$

where μ is the chemical potential, n_i the occupation number where $n_i=1$ if site i is occupied and zero otherwise. The prime on summations means the exclusion of the term with $i=j$. J_{ij} describes the interaction between adatoms at sites i and j: $J_{ij}=g/|\vec{r}_i-\vec{r}_j|^{2\sigma}$, where g and σ are constants. For the dipole-dipole interaction $\sigma=3/2$. $\vec{r}_i$ is the position vector of the th site.

The system seems very difficult to investigate because of the nature of repulsive long-range interaction. Nevertheless, it has a remarkable advantage in that an ordered structure inherent in the intrinsic interaction is formed. This recognition prompts us to do the present work.

2. Structures on a Smooth Surface

As a preliminary for finding ordered structures in the lattice gas model, we consider a dipole gas on a smooth substrate where dipoles take any positions. We assume adatoms form a Bravais lattice on a smooth surface. The positions of the Bravais lattice are given by vectors as $\vec{r}=m\vec{a}+n\vec{b}$, where $\vec{a}$ and $\vec{b}$ are primitive vectors, and m and n are any integers. The $\vec{a}$, $\vec{b}$, and $\vec{c}$ $(=\vec{b}-\vec{a})$ form a triangle, and the following restrictions may be imposed without loss of generality:

$$a \leqq b \leqq c \;, \qquad 0 \leqq x_a,\ x_b,\ x_c \leqq 1/2, \tag{2}$$

where a, b, and c stand for the length of sides of the triangle, and x_a, x_b, and x_c are defined as $x_a=(\vec{a},\vec{b})/a^2$, $x_b=(\vec{b},\vec{a})/b^2$, and $x_c=-(\vec{c},\vec{a})/c^2$. Moreover the area of the primitive cell is assumed to be unity if one is not concerned with the absolute value of the interaction.

The purpose of this section is to give a condition deciding which is more energetically favourable between any two Bravais lattices. The interaction energy in eq.(1) can now be written by using these notations as

$$E_i = \frac{1}{2}N\theta\varepsilon(a,x_a), \qquad \varepsilon(a,x_a) = g\sum_{mn}{}'[a^2(m+nx_a)^2+n^2/a^2]^{-\sigma}, \tag{3}$$

where θ is the coverage, and the $\varepsilon(a,x_a)$ is the interaction energy per atom. The prime on summations means exclusion of the term with m=n=0. The $\varepsilon(a,x_a)$ can be also regarded as a function of b and x_b, or c and x_c. It can be shown that the function $\varepsilon(z,x_z)$ ($z=a,b$, or c) has the following property:

$$\left.\frac{\partial\varepsilon(z,x_z)}{\partial x_z}\right|_z \leqq 0, \tag{4}$$

where the sign of equality holds only if x_a, x_b and x_c take 0 and 1/2, respectively. This immediately means that the dipole gas is unstable against the shear deformation parallel to a fixed side of the triangle as far as the vertex is not on the bisector perpendicular to that side. From this result it is concluded that the dipole-dipole interaction drives adatoms to form a regular triangular lattice. This agrees with results recently obtained by molecular dynamic technique [5] that the system crystallizes into a regular triangular lattice due to repulsive dipole-dipole interaction. It is worthwhile noting that the inequality (4) holds for a wide class of repulsive interaction including Coulomb interaction.

The triangle can be also specified by a, b and c instead of a and x_a, or b and x_b, or c and x_c. By using a, b, and c, a set of inequalities (4) is united in the one inequality [4]:

$$\varepsilon(a',c') \leqq \varepsilon(a,c) \qquad \textit{if } a \leqq a' \textit{ and } c' \leqq c, \tag{5}$$

where $a'\leqq b'\leqq c'$ and $a\leqq b\leqq c$ are imposed. We note that no condition is imposed between b and b' in (5).

Two cases of $a<a'$ and $c<c'$, $a'<a$ and $c'<c$ remain untouched. However, both cases become equivalent if one reads a and c as a' and c', respectively; so we hereafter consider only the former case. For this case, we have the following inequality [4]:

$$\varepsilon(a',c') \leq \varepsilon(a,c) \qquad\qquad \textit{if } a < \gamma a' \textit{ and } c < c', \tag{6}$$

where γ is defined as the ratio, $\gamma = a/a'$, a constant given by γ=0.9485..., where a and a' satisfy $\varepsilon(a',x_a'=0)=\varepsilon(a,x_a=1/2)$. It should be noticed here that γ is very close to unity. For this reason, an untouched case, $\gamma a'<a<a'$ and $c<c'$, is quite limited. The deviation of γ from unity represents the degree of the stability of the lattice of the adlayer against the shear defomation. We can point out that the value of γ decreases as the exponent of the interaction, σ, increases: for Coulomb system γ becomes closer to unity. The value of γ does not depend on the assumed area of the primitive cell.

The case that the inequalities (5) and (6) leave untouched is that of $\gamma a'<a<a'$ and $c<c'$. For this case, there is no simple rule like (5), and (6) about which structures specified by two different triangles lower the interaction energy. It is necessary to directly calculate the interaction energy for each of two assumed structures and to compare them, in order to see which structure is more favourable.

3. Commensurate Structures

We are now in a position to present the method for constructing commensurate structures of the adlayer of adatoms on various types of substrate. Only the simplest case of coverage of $\theta=1/p$, where p is an integer, is considered.

Let the primitive vectors of the substrate lattice be denoted by $\vec{g}_1$ and $\vec{g}_2$. Without loss of generality the area of the primitive cell is assumed to be unity: $|\vec{g}_1 \times \vec{g}_2| = 1$. Again let $\vec{a}$ and $\vec{b}$ denote the primitive vectors of the lattice of adatoms in accord with the definition in §2, where $\vec{a}$ and $\vec{b}$ obey the conditions (2). The vectors $\vec{a}$ and $\vec{b}$ can be written in terms of those of the substrate lattice as

$$\vec{a} = m\vec{g}_1 + n\vec{g}_2 \qquad \text{and } \vec{b} = m'\vec{g}_1 + n'\vec{g}_2,$$

where m, n, m' and n' are integers.

The problem now is to find a set of integers m, n and m', n'. POKROVSKY and UIMIN [6] have carried this out by means of a direct minimization of the interaction energy for all possible configurations of the adlayer. We take a different approach.

We have seen in §2 that the intrinsic interaction drives adatoms to form a favourable structure. For the real system, where adatoms take discrete points, this tendency still holds. We explore a favourable set of integers, based on results obtained in §2. We note that a set of integers (m,n) and (m',n') is chosen among sets of (m,n) and (m',n') where
(*Ia*) a set of integers (m,n) satisfies the inequality:

$$(\vec{a},\vec{a}) \equiv m^2|\vec{g}_1|^2 + 2mn(\vec{g}_1,\vec{g}_2) + n^2|\vec{g}_2|^2 \leqq 2p/\sqrt{3}, \tag{7}$$

(*Ib*) a set of integers (m',n'), satisfying $|mn'-m'n|=p$, exists, and

(*Ic*) a set of integers (m',n') maximizes $x_a \equiv(\vec{a},\vec{b})/(\vec{a},\vec{a})$.

(*Ia*) comes from the condition (2) which says that the upper limit of the length of the shortest side of the triangle is bounded by that of the side of the regular triangle, (*Ib*) from the condition $|\vec{a}\times\vec{b}|=p$, and (*Ic*) from the property of the interaction energy (3).

Our method for finding commensurate structures of adlayer for a given integer p is as follows.

(*IIa*) *We choose a set of integers* (m_1,n_1) *defining the longest length,* a_1, *of a among sets of integers* (m,n) *satisfying* (*Ia-c*).

This procedure gives a triangle. Let c_1 denote the longest length of this triangle, and let k (≥ 2) be any positive integer, and start from 2.

(*IIb*) *We try to choose a set of integers* (m_k,n_k) *defining the k-th longest length,* a_k, *of a and its corresponding length,* c_k, *of c such that the* a_k *and* c_k *satisfy:*

$$\gamma a_{k-1} < a_k < a_{k-1} \quad \text{and } c_{k-1} < c_k, \text{ and } \varepsilon(a_k,c_k) < \varepsilon(a_{k-1},c_{k-1}). \tag{8}$$

If one fails to choose a set of integers (m_2,n_2) in (*IIb*), the favourable set is given by (m_1,n_1). This is the outcome from the inequalities (5) and (6). If one succeeds, one needs to repeat the procedure (*IIb*) by increasing k one by one so as to find a new set of integers (m_3,n_3) and so on as far as the sequence ends in a set of integers, say (m_k,n_k), satisfies $a_k > \gamma a_1$. Then, the favourable set is (m_k,n_k).

Sets of integers (m,n) and (m',n') giving favourable structures up to p=30 are given in Table I for the square and the regular triangular lattice of the substrate. For most cases the set of (m,n) determined from (*IIa*) gives the most favourable structure. Exceptions, for example, appear for p=26, 27, 40 and 50 for the square lattice and for p=45 for th regular

square lattice			regular triangular lattice		
p	$\bar{a}$ (m, n)	$\bar{b}$ (m',n')	p	$\bar{a}$ (m, n)	$\bar{b}$ (m',n')
1	(1, 0)	(0, 1)	1#	(1, 0)	(0, 1)
2	(1,-1)	(1, 1)	2	(1, 0)	(0, 2)
3	(1,-1)	(2, 1)	3#	(2,-1)	(1, 1)
4	(2, 0)	(1, 2)	4#	(2, 0)	(0, 2)
5	(2,-1)	(1, 2)	5	(2,-1)	(1, 2)
6	(2,-1)	(2, 2)	6	(2, 0)	(2,-3)
7	(2,-1)	(-1,-3)	7#	(3,-1)	(1, 2)
8	(2,-2)	(3, 1)	8	(3,-1)	(1,-3)
9	(3, 0)	(2, 3)	9#	(3, 0)	(0, 3)
10	(3,-1)	(1, 3)	10	(3,-1)	(1, 3)
11	(3,-1)	(2, 3)	11	(3,-1)	(1,-4)
12	(3,-2)	(3, 2)	12#	(4,-2)	(2, 2)
13	(3,-2)	(2, 3)	13#	(4,-1)	(1, 3)
14	(3,-2)	(4, 2)	14	(4,-2)	(1, 3)
15	(4,-1)	(3, 3)	15	(4,-1)	(-1, 4)
16	(4, 0)	(2, 4)	16#	(4, 0)	(0, 4)
17	(4,-1)	(1, 4)	17	(4,-1)	(3,-5)
18	(3,-3)	(4, 2)	18	(4,-1)	(2,-5)
19	(4,-1)	(3, 4)	19#	(5,-2)	(2, 3)
20	(4,-2)	(4, 3)	20	(4, 0)	(4,-5)
21	(3,-3)	(5, 2)	21#	(5,-1)	(1, 4)
22	(4,-2)	(-1,-5)	22	(5,-2)	(1, 4)
23	(4,-3)	(5, 2)	23	(4,-1)	(-1, 6)
24	(5,-1)	(1,-5)	24	(5,-1)	(-1, 5)
25	(4,-3)	(3, 4)	25#	(5, 0)	(0, 5)
26†	(4,-3)	(-2,-5)	26	(5,-1)	(4,-6)
27†	(4,-3)	(5, 3)	27#	(6,-3)	(3, 3)
28	(5,-2)	(4, 4)	28#	(6,-2)	(2, 4)
29	(5,-2)	(2, 5)	29	(5,-1)	(-1, 6)
30	(5,-3)	(5, 3)	30	(6,-3)	(2, 4)

Table I

Sets of integers (m,n) and (m',n') specifying primitive vectors of commensurate structures of the adlayer for the square and the triangular lattice of the substrate. The coverage is given by $1/p$ (p: integer).

triangular lattice. Results coincide with those by POKROVSKY and UIMIN, who considered only the case of $p \leq 21$ for the regular triangular lattice of the substrate.

4. Interaction Energy

We are here concerned with the ground-state energy, to give its lower bound for a whole range of coverage. As is shown in §2, adatoms form a regular triangular lattice due to the repulsive long-range interaction, while its tendency is suppressed by the substrate, so that the interaction energy increases more than that of the adlayer with the regular triangular structure. So the lower bound for the ground-state energy may be evaluated by assuming adatoms form the regular triangular lattice. We first confine ourselves to the case of $\theta \leq 1/2$. One has

$$E_0 = N\theta\varepsilon_0(\theta),$$
$$\varepsilon_0(\theta) = \mu + \frac{1}{2}\theta^{\sigma}F_0, \qquad 0 \leq \theta \leq 1/2, \tag{9}$$

where F_0 is defined as the interaction energy per atom when adatoms occupy entirely the regular triangular lattice.
For the case of $1/2 \leq \theta$, the particle-hole symmetry is used to give a lower bound for the ground-state energy, yielding

$$E_0 = N\theta\varepsilon_0(\theta),$$
$$\varepsilon_0(\theta) = \mu + \frac{1}{2\theta}[F_s(2\theta-1) + (1-\theta)^{1+\sigma}F_0], \quad 1/2 \leq \theta \leq 1. \tag{10}$$

The coverage-dependence of the energy is indicated in Fig.1 by solid lines for the cases of the square and regular triangular lattices of the substrate, assuming $\mu=0$. The energy given in eq.(12) and (13) agrees very well with true values indicated by black and white circles, except for near one half of the coverage. The reason for this good agreement for the case of the low and high coverage is explained as follows.

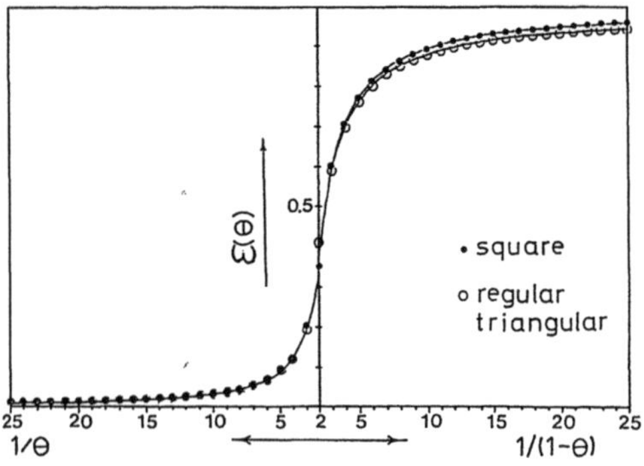

Fig. 1

The lower bound for the ground-state energy per atom as a function of $1/\theta$ for the cases of the square and triangular lattices of the substrate, indicated by solid lines assuming $\mu=0$.

Consider the case that the coverage is very low. The coverage imposes the condition $|mn'-m'n|=p$ on sets of integers (m,n) and (m',n') of vectors $\vec{a}$ and $\vec{b}$. In the very low coverage, p becomes large. As p becomes larger and larger, the number of sets of integers (m,n) and (m',n') satisfying $|mn'-m'n|=p$ may increase more and more. Among these sets vectors $\vec{a}$ and $\vec{b}$ are chosen so as to form a regular triangular lattice as far as possible according to the discussions in §3, and then the triangle spanned by $\vec{a}$ and $\vec{b}$ becomes closer and closer to the regular one. This reasoning can be applied to the case of the high coverage if one uses the particle-hole symmetry.

5. Discussion and Conclusion

The basic ingredients of this work are as follows. The adatoms are forced to form an ordered structure inherent in the intrinsic interaction on the substrate. But its tendency is suppressed by a strong binding by the substrate, since adatoms are not allowed to take arbitrary positions. The misfit between the ordered structure of adatoms due to intrinsic interaction and the substrate structure is nothing but the frustration [6].

Let us go back to the inequality (7). The sign of equality holds only if adatoms are allowed to form a regular triangular lattice. The frustration always takes place if no common measure exists, except a rational number between $(\vec{g}_i,\vec{g}_j)$ and $2/\sqrt{3}$ where $i,j=1,2$. For the case where the rational common measure exists, (7) becomes equivalent to the Diophantine equation in a quadrutic form [7], which can be examined on the basis of those of Pell's equation.

For the square lattice of the substrate where $(\vec{g}_i,\vec{g}_i)=1$ and $(\vec{g}_1,\vec{g}_2)=0$, there is no rational common measure. Thus the commensurate structures are frustrated for the coverage of $1/p$. It can be proved that this is also true for the whole range of coverage. In contrast to this, the regular triangular lattice of the substrate has the possibility of leading unfrustrated lattice of adatoms. For a class of coverage given by $\theta=1/p_{mn}$, where $p_{mn}=m^2+mn+n^2$, m and n being any positive integers, adatoms are allowed to form a regular triangular lattice since $(\vec{g}_i,\vec{g}_i)=2/\sqrt{3}$ and $(\vec{g}_1,\vec{g}_2)=1/\sqrt{3}$. Sets of integers (m,n) and (m',n') representing the unfrustrated structures are indicated in Table by symbol (*).

References

1. L. A. Bol'shov, A. P. Napartovich, A. G. N. Naumovets and A. G. Fedorus: Sov. Phys. Usp. 20 (1977) 432
2. M. Kaburagi and J. Kanamori: J. J. Appl. Phys. Suppl. 2 (1774) 145
3. V. L. Pokrovsky and G. V. Uimin: J. Phys. C, Solid St. Phys. 11 (1978) 3535
4. K. Shinjo and T. Sasada: J. Phys. Soc. Jpn. (submitted)
5. P. Vashishta and R. K. Kalia: Melting, Localization, and Chaos ed. R. K. Kalia and P. Vashishta (Elsevier Sci. Publishing Co, Inc. 1982) p.43
6. G. Toulouse: Commun. on Physics 2 (1977) 115
7. G. H. Hardy and E. M. Wright: Introduction to the Theory of Numbers

Temperature Effect in Surface Reconstruction of Clean and Hydrogen-Adsorbed W(001) Surface

T. Matsubara
Department of Physics, Kyoto University, Kyoto 606, Japan

1. Introduction

It is now well-known that the clean W(001) surface undergoes a reconstructive phase-transition at certain temperature T_c and exhibits a c(2×2) structure at low temperatures. From various experimental and theoretical sources, it is concluded that this surface reconstruction is a second order transition caused by spontaneous displacements of surface atoms along a direction parallel to the surface, as depicted in Fig.1 [1]. The reconstructed structure of W(001) is somehow stabilized when hydrogen atoms are adsorbed, so long as the coverage of hydrogen is small [2]. As to the mechanism of the phase-transition, several models have been proposed [3], but at present it seems that an accepted view point is to attribute the origin of driving transition to the competition between stabilizing band electronic force and opposing ion-core repulsive force. Among others, TERAKURA et al. gave an electronic theory of the surface reconstruction on W(001) [4] which could explain even quantitatively the relevant experimental facts for absolute zero T=0. As to temperature-variation of this transition, however, there are a few theories in which the temperature-effect is seriously considered. Probably this is due to the fact that the main source of information on the temperature-dependence of the "order parameter" is LEED experiments, which are much influenced by large surface vibration as far as the temperature-effect is concerned. Indeed, for instance, we can observe in Fig.2, with increasing temperature, strong exponential decay of the scattering intensity of the LEED patterns characteristic to the low-temperature phase [5]. If we look at Fig.2 as a temperature-dependence curve for the "order parameter" of the phase-transition, it would be quite unusual in contrast

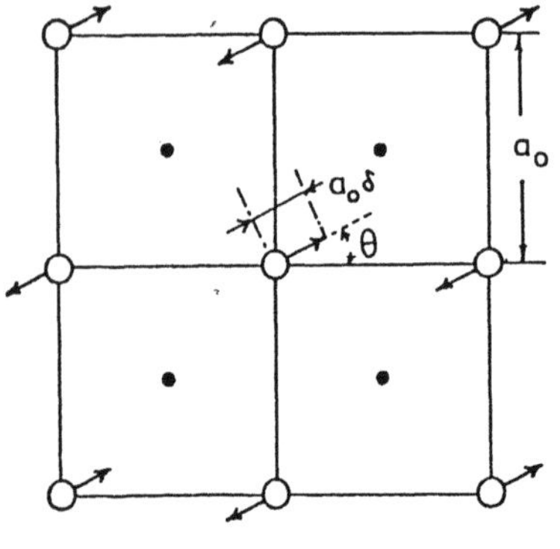

Fig.1 Atomic displacements along the reconstructed W(001) surface. The white circles represent the surface atoms and the black points the second layer atoms. The stable configuration is $\Theta=\pi/4$.

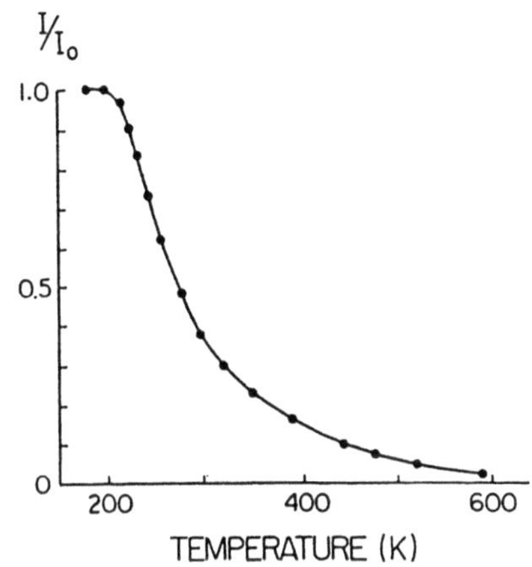

Fig.2 Temperature dependence of the intensity of diffuse spots in LEED pattern from the reconstructed W(001) surface

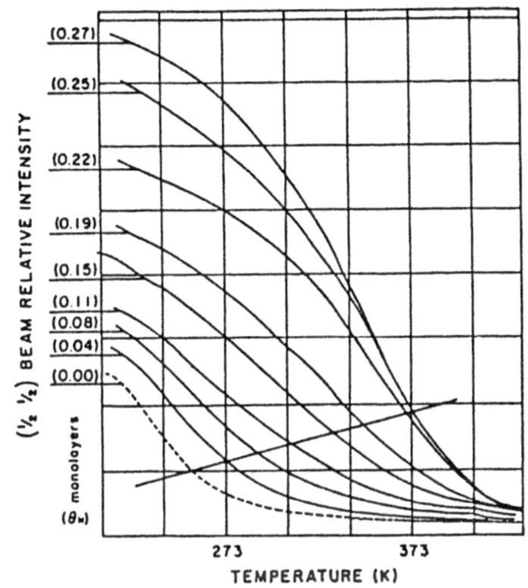

Fig.3 Intensity of the $(\frac{1}{2},\frac{1}{2})$ beam as a function of temperature at several coverages [6]. The coverage is shown in the left.

to the standard behaviour of the order parameter in the second order phase-transition: The transition-temperature is obscured by the exponential tail.

This trend is even enhanced in the case of hydrogen-adsorbed surface as demonstrated in Fig.3, which is reproduced from the original paper by BARKER and ESTRUP [6]. According to the authors, they determine the transition-temperature for each coverage by "drawing arbitrarily a curve" as indicated in the figure and tracing its cross points with the LEED intensity curves. It would be very hard to give a definite answer for T_C by such procedure.

One way to make the transition point sharp is, as suggested by KING [7], to utilize the data of LEED scattering of integral order corresponding to the ordinary LAUE-BRAGG LEED spots, instead of the scattering intensity data of half integral order corresponding to the super structure spots. Fig.4 shows some examples for the LEED scattering intensities of integral order plotted as functions of temperature in semi-log scale for different incident electron energies. As seen from the figure, at high temperatures all curves become beautifully linear in T, indicating the Debye-Waller type behaviour. Indeed, these are consistent with the well-known formula for the scattering intensity

$$I = I_0 \exp[-W(T)] \qquad (1.1a)$$

where W(T) is the Debye-Waller function defined by

$$W(T) = \langle(\vec{k}\cdot\vec{x})^2\rangle \cong (12h^2/Mk_B)(2\pi/\lambda)^2(T/\Theta^2) \qquad (1.1b)$$

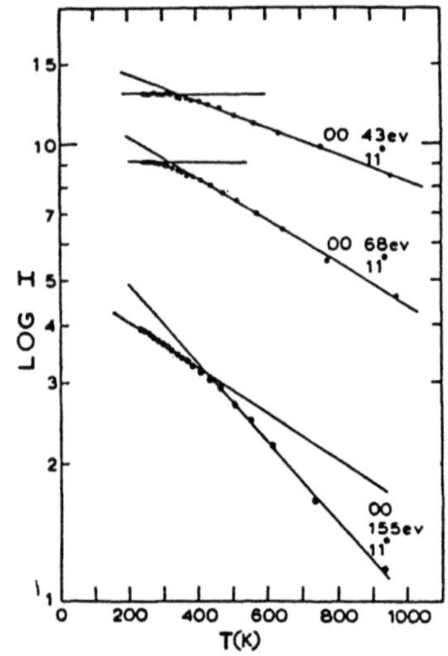

Fig.4 Log intensity versus temperature plots for several peaks of the (00) spectra. The change in slope takes place around T_C=400K. [7]

M the atomic mass of W, $\vec{x}$ displacement vector of thermal vibration, Θ Debye temperature and the square of the electron wave vector

$$\vec{k}^2 = (2\pi/\lambda)^2 = 2mE/h^2 \qquad (1.2)$$

is related with the electron energy E as in (1.2). It is easily checked that all the curves in Fig.4 satisfy (1.1) with (1.2) at high temperatures and thus we can reduce from them

$$\Theta \simeq 200\text{K} \qquad (1.3)$$

which is the surface Debye temperature much smaller than the bulk value Θ(bulk)=310K.

In addition to this high temperature linear dependence, we can clearly observe a kink in each curve at around T_C=400K, below which some part of the scattering intensity is lost. If this deficient intensity has to be taken over by the scattering of half-integral order, which starts to appear with the surface reconstruction, then T_C may be identified with the transition temperature.

In order to substantiate such an argument, we propose in this paper a theory of surface reconstruction of W(001) at finite temperature and try to simulate theoretically the observed data such as those involved in Fig.4.

2. Theory

Let us describe the reconstructed c(2×2) structure by two penetrating 2D sub-lattices and let $\vec{u}$ be the displacement vector of one sub-lattice and $-\vec{u}$ that of another sub-lattice (see Fig.1). First we shall point out that an important hint may be extracted from Fig.4 on the temperature-dependence of the displacement $\vec{u}$. It is not difficult to prove that the LEED intensity for integral order is given by

$$I = I(H,K)\ \exp[-W(T)]\ \cos^2(\vec{k}\cdot\vec{u})$$

$$=)\ (H.K)\ \exp[-W(T)]\,[1-(\vec{k}\cdot\vec{u})^2] \qquad (2.1a)$$

and for half integral order

$$I = I(H+\tfrac{1}{2},\ K+\tfrac{1}{2})\ \exp[-W(T)]\sin^2(\vec{k}\cdot\vec{u})$$

$$= I(H+\tfrac{1}{2},\ K+\tfrac{1}{2})\ \exp[-W(T)]\,(\vec{k}\cdot\vec{u})^2 \qquad (2.1b)$$

where $\vec{k}$ is the scattering electron wave vector, and (H,K) are a pair of integers designating a lattice point in the 2D reciprocal vector space. We note in Fig.4 that the deflection angles formed by two straight lines at kink are almost identical for different electron energies. This strongly suggests that u^2 is also an exponential function of T. Keeping this point in mind, we shall construct a free-energy functional F(u) by extending Terakura's theory for T=0 to finite temperature.

To begin with, let us summarize briefly the theory for T=0. Introducing the parameters $\delta=|u|\ a_0$ and θ defined in Fig.1, Terakura et al. calculated the energy of the system with the reconstructed surface as a function of δ and θ in the form:

$$\Delta E(\delta,\theta) = A_1\delta^2 + (A_2 + A_3\sin^2 2\theta)\,\delta^4 + \cdots \qquad (2.2)$$

Each coefficient consists of two contributions: one from the band electron (suffix B) and the other from atomic core repulsion (suffix R):

$$A_n = A_{n,B} + A_{n,R} \qquad n = 1,2,3. \qquad (2.3)$$

The band electron contributions have been evaluated by means of the recursive method, using the standard d-band parameters for W, and the repulsive parts have been derived by making use of the simple Born-Mayer potential

$$\Delta E_R = \sum_{ij} B\ \exp[-p(R_i - R_j)] \qquad (2.4)$$

with

$$p = 2.05/a_H\ \ (a_H\text{=Bohr radius})\ \text{and}\ B=2665\text{Ryd}. \qquad (2.5)$$

The numerical values for the A's in Ryd are:

$$\begin{array}{lll} A_{1R} = 9.0686 & A_{1B} = -9.7813 & A_1 = -0.7127 \\ A_{2R} = 180.537 & A_{2B} = -88.968 & A_2 = 91.869 \\ A_{3R} = -93.563 & A_{3B} = 66.726 & A_3 = -26.837. \end{array} \qquad (2.6)$$

From (2.6) it is clear that $\theta=\pi/4$ is the most stable configuration and the electronic band energy makes the reconstructed structure more stable, while the repulsive forces act to prevent it.

The nature of the phase-transition associated with the surface reconstruction is essentially similar to that of the so-called displacive transition in ferroelectrics: The symmetric structure of high-temperature phase is stabilized by non-linear anharmonic vibrations, and with decreasing temperature, the decrease in anharmonicity makes the symmetric structure less stable eventually resulting in the transition to low-symmetry structure with lower energy. In order to discuss such temperature effect, we apply the "self-consistent Einstein model (SCEM)" developed by us for other purpose.[8]

In SCEM each atom is assumed to vibrate independently in each average potential, and it is generally proved that when an atom at R executes thermal vibration with amplitude x, the thermal average of the potential of Eq.(2.4) is given by

$$\langle \exp[-p(R+x)]\rangle_x = \exp(-pR)\exp[\tfrac{1}{2}W_p(T)] \qquad (2.7)$$

where

$$W_p(T) = p^2\langle x^2\rangle = (h^2p^2/2Mk_B\Theta)\coth(\Theta/2T) \qquad (2.8)$$

and Θ is the frequency of the Einstein oscillator expressed in temperature unit. Note the difference between the present average and the case of Debye-Waller function: The sign in the exponent is oppisite and k^2 is replaced by p^2. At high temperature, (2.7) gives an exponentially increasing function of T. It is a straightforward task to include the temperature-dependence in the coefficients A_{nR}'s once the average interatomic repulsive potential is given. In the present case, it is just to differentiate the functions in (2.4) with respect to R_i and to take thermal average of (2.7). Thus it turns out that a common factor $\exp[W_p(T)]$ appears in all the terms of ΔE_B. Taking $0=\pi/4$, and putting $A_{2v}+A_{3v}=A^*_{2v}$ (v=B or R), we obtain the free energy functional F(u), the extension of $\Delta E(\delta,\pi/4)$ for finite temperature, as

$$F(u) = A_1(T)\,\delta^2 + A^*_2(T)\,\delta^4 + \cdots \qquad (2.9)$$

with in Ryd

$$A_1(T) = -9.7813 + 9.0686\,\exp[W_p(T)] \qquad (2.10a)$$

$$A^*_2(T) = -22.242 + 87.274\,\exp[W_p(T)] \qquad (2.10b)$$

It is obvious that the equilibrium displacement $u_0=\delta a_0$, which makes F(u) minimum, is finite at low temperature but vanishes at and above T_c determined from

$$A_1(T_c) = 0 \qquad (2.11a)$$

or using (2.8)

$$\tanh(\Theta/2T_c) = (h^2p^2/2Mk_B\Theta)/\ln(9.7813/9.0686) \qquad (2.11b)$$

As expected from experiments, $u_0(T)$ for $T<T_c$ is shown to vary exponentially and furthermore the equation (2.11b) gives for Θ=200K a critical temperature T_c=450k, a value of right order. Thus our theory seems to describe the problem quite well.

3. Comparison with Experiments

We are now in a position to test our theory on a more quantitative basis. For this purpose, it will be interesting if we can work out from our theory some results which may be compared directly with some relevant experiments. For instance we shall focus our attention on Fig.4 and try to calculate the log intensity versus temperature relations of (00) spectra for the precisely same condition as the real experiments so that one-to-one correspondence between theoretical results and experimental observations may be made directly.

Thus, what we have to calculate is, using (2.1a),

$$\ln[I/I(00)] = -W(T) + \ln[1-(\vec{k}\cdot\vec{u})^2] \tag{3.1}$$

In order to be consistent with our SCEM, we first rewrite W(T) in terms of Θ, the Einstein temperature, as

$$\begin{aligned} W(T) &= (h^2k^2/2Mk_B\Theta)\coth(\Theta/2T) \\ &= (m/M)(E/k_B\Theta)\coth(\Theta/2T) \end{aligned} \tag{3.2}$$

In the second line of (3.2), use has been made of (1.2), and m denotes the electron mass. For the seocnd term in the right hand side of (3.1), again upon using (1.2) we may put

$$(\vec{k}\cdot\vec{u})^2 = f(2mE/h^2)u_0^2(T) \tag{3.3}$$

where $u_0^2(T)$ is the equilibrium atomic displacement given by

$$[u_0(T)/a_0]^2 = \begin{cases} -A_1(T)/2A_2^*(T) & \text{for } T < T_C \\ 0 & \text{for } T > T_C \end{cases} \tag{3.4}$$

and f is a factor of order-unity depending on the direction of $\vec{k}$. Inserting both (3.2) and (3.3) in (3.1), we have a basic formula to calculate the log intensity as a function of temperature. There are few adjustable parameters: The force range parameter p in Born-Mayer potential has been fixed by TERAKURA et al. [4] from the experimental data on the lattice constant and compressibility of the bulk metal. Only one adjustable parameter involved here is Θ. Strictly speaking, Θ has also to be determined self-consistently in the spirit of SCEM, although we have omitted this procedure, leaving Θ as a parameter.

We have performed numerical simulations based on the above formulation for log [I/I(00)], taking the input parameters

$$E = 155\text{eV}, \quad 68\text{eV}, \quad 43\text{eV}$$

corresponding to the experiments in Fig.4, and an adjustable parameter Θ=200K. The result is shown in Fig.5 where the temperature-variation of the order parameter $\delta^2=[u_0(T)/a_0]^2$ is also included. It is quite gratifying that the numerical simulation reproduces most of the characteristic features of the experiments.

4. Discussion

The theory developed in the previous sections can be easily extended to the case of hydrogen-adsorbed surface. As seen in Fig.3, the hydrogen-adsorbed W(001) surface also exhibits phase-transitions similar to that of the clean surface, but at higher T_C with increasing hydrogen coverage. This effect has been explained by LAU and YING[9], using a simple model.

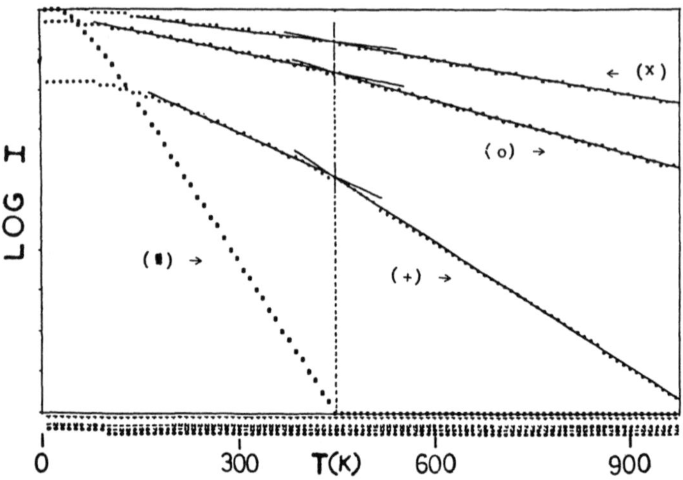

Fig.5
Numerical simulation of log-intensity versus temperature relations corresponding to Fig.4. An adjustable parameter is taken as Θ=200K.
▮: $\delta^2(T)$, +:E=155eV, o:E=68eV, x:E=43eV

They assume that each adsorbed hydrogen atom occupies a bridge site (a midpoint between two neighbouring surface W atoms), and introduces the hydrogen occupation number n_i at site i so as to describe the adsorbed system by a Hamiltonian of the form:

$$H_{ad} = \sum_{ij} J_{ij} n_i n_j + \sum_{is} \vec{\lambda}(i-s) n_i \vec{u}_s \qquad (4.1)$$

J_{ij}'s denote the H-H interactions and $\vec{\lambda}(i-s)$ involves the gradient of the H-W interaction potential between i H-site and s W-site. Here, general atomic displacements u_s is introduced depending on s and therefore the Hamiltonian (4.1) may describe not only the homogeneous distortion $\vec{u}_0$ but also more complicated incommensurate distortions as discussed by INAOKA and YOSHIMORI [10]. Within the assumption of homogeneous distortion, if one performs the calculation of the partial free-energy from $\mathrm{Tr}[\exp(-H_{ad}/k_BT)]$ with fixed $\vec{u}$ and eliminates the freedom of hydrogen by minimizing the partial free-energy, then it turns out that, irrespective of the sign of the H-W potential, a negative definite quadratic form $C(T)\delta^2$ $(C(T)<0)$ is left as a term to be added to the free-energy functional F(u) of the clean surface system, and hence the effective free-energy functional for the hydrogen-adsorbed surface becomes:

$$F_{eff}(u) = [A_1(T) + C(T)]\delta^2 + A_2^*(T)\delta^4 + \cdots \qquad (4.2)$$

This means that the coupling of the H adatoms to the substrate distortion leads to an enhancement of the displacive transition as observed in Fig.3. LAU and YING have treated the displacive transition of the clean surface phenomenologically, but when their phenomenological part is substituted by our microscopic theory, an essentially same result will come out.

So far, we have neglected the temperature-effect in the band electron energies. In general, however, this may not be justified. Indeed, it is known that the lattice spacing of W near surface shows a contraction of 6% and, as mentioned already, a remarkable softening of surface phonon takes place, as indicated by a change in Θ(bulk)=310K to Θ(surface)=200K. These effects cannot be attributed only to the anharmonicity, and should be discussed on the same footing as the surface reconstruction. Thus, the best approach would be such that the changes in both electron and phonon states due to the presence of the surface (and adatoms) were treated in self-consistent way. Based on such consideration, we are led to a new scheme for studying surface problems, including surface relaxation, surface reconstruction and surface softening at the same time, a scheme which is a combination of the SCEM for phonon system and a "dynamical CPA" for electron system (MATSUBARA and ODAGAKI [11]). We hope to report an application of this new fomulation to the present problem elswhere.

References

1. D.A.King:Physica Scripta T4,34(1983).
2. M.K.Debe and D.A.King:J.Phys.C10,L1303(1977).
   T.E.Felter,R.A.Barker and P.J.Estrup:Phys.Rev.Lett.38,1138 (1977).
3. J.E.Inglesfields:Vacuum 31,663(1981).
4. K.Terakura,I.Terakura and Y.Teraoka:Surface Sci.80,535(1979).
   I.Terakura,K.Terakura and N.Hamada:Surface Sci.103,103(1981).
   K.Masuda-Jindo,N.Hamada and K.Terakuka:J.Phys.C17,L271(1984).
5. D.A.King and G.Thomas:Surface Sci.72,201(1980)
6. R.A.Barker and P.J.Estrup:J.Chem.Phys.74,1442(1981)
7. M.K.Debe and D.A.King:Surface Sci.81,193(1979)
8. T.Matsubara and K.Kamiya:Prog.Theor.Phys.58,767(1977).
   T.Matsubara,Y.Iwase and A.Momokita:Prog.Theor.Phys.59,1102 (1978).
   T.Hama and T.Matsubara:Prog.Theor.Phys.59,1407(1978).
9. K.H.Lau and S.C.Ying:Phys.Rev.Lett.44,222(1980)
10. T.Inaoka and A.Yoshimori:Surface Sci.115,301(1982).
11. T.Matsubara and T.Odagaki:in preparation.

Phenomenological Analysis of the W(001) Reconstructed Surface

T. Inaoka, M. Sawada*, and A. Yoshimori

Department of Material Physics, Faculty of Engineering Science
Osaka University, Toyonaka, Osaka 560, Japan

1. Introduction

Recently the W(001) reconstructed surface and the effects of adatoms on the reconstruction have been studied intensively. These studies show that the surface displacement modes (the M_5 mode and the incommensurate modes around the M point in the 1st Brillouin zone) have instability towards surface reconstruction. By means of a lattice dynamical approach and a symmetry consideration of the system we investigate the eigenvectors of these displacement modes and the energy dispersion, and present the general form of the Ginzburg-Landau (G-L)-type free energy functional describing the reconstructed surface. In the course of this study, we discuss also the Mo(001) reconstructed surface. On the W(001) reconstructed surface, the interaction between adatoms and "displaced" surface W atoms might make a major contribution to the adsorbate-induced changes of surface W atom displacements. We examine the effects of this interaction on the commensurate - incommensurate (C-IC) displacive transition, that is, the instability of the C phase towards the IC phase.

2. Surface Phonons Leading to Reconstruction

We take a simple approach that W atoms in a few surface layers are allowed to vibrate and those in deeper layers are fixed at their lattice positions. The equations of motion describing this surface vibration are reduced to a matrix (dynamical matrix) eigenvalue problem for each of the 2-dimensional wave vectors. We take account of the nearest and next-nearest interatomic interaction given by a pairwise central force potential. By varying two kinds of the surface-nearest-neighbor force constants (following FASOLINO, SANTORO and TOSATTI [1]), we investigate which mode has instability towards reconstruction. Though this approach (the three surface layers are allowed to vibrate.) is very simple, we can reproduce the result obtained by means of a slab calculation [1]. Next, we apply the same approach to the Mo(001) surface. We can obtain the IC Σ phase observed on the clean surface (Fig. 1 and [2]) by varying one nearest-neighbor force constant between the first and second layers, in addition to the above-mentioned force constants. It might suggest the effect of relaxation on the reconstruction.

These analyses have been done by solving numerically the eigenvalue problem of the surface phonon modes. Next, we focus our attention on the surface phonon modes having wave vectors near the M point in the 1st Brillouin zone. We treat the dynamical matrix up to the second order in k_x or k_y, where wave vector $\vec{k}$ $(=(k_x,k_y))$ is measured from the M point (Fig. 1). There are two degenerate modes at the M point (M_5 mode [1]) which lead to the Debe-King model for the clean surface reconstruction [3]. The two dispersion branches,

*Present address: Research Center, Sanyo Electric Co., Ltd
Hirakata, Osaka 573, Japan

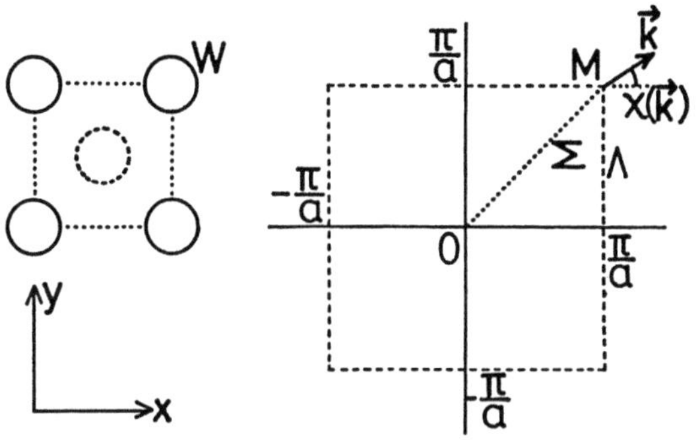

Fig. 1 Unit cell on the surface and the 1st Brillouin zone of the square lattice. The broken circle represents a second-layer W atom

which start from the M_5 mode, split for finite $\vec{k}$ values. The displacive eigenvectors of the first layer with $\vec{k}$ on these two branches are given by

$$\vec{u}_1(\vec{k}) = (1, 0, i\zeta k_x)^t \cos\vartheta(\vec{k}) + (0, 1, i\zeta k_y)^t \sin\vartheta(\vec{k}) \qquad (1)$$

and

$$\vec{u}_2(\vec{k}) = -(1, 0, i\zeta k_x)\sin\vartheta(\vec{k}) + (0, 1, i\zeta k_y)^t \cos\vartheta(\vec{k}) \qquad (2)$$

where ζ is a real constant, t represents transposition of the vector, and $(\cos\vartheta(\vec{k}), \sin\vartheta(\vec{k}))^t$ is the eigenvector for the lower eigenvalue of the matrix of the form

$$\begin{pmatrix} Ak_x^2 + Bk_y^2 & Ck_xk_y \\ Ck_xk_y & Bk_x^2 + Ak_y^2 \end{pmatrix} \qquad (3)$$

with real constants, A,B and C. The phase of the z-components of these eigenvectors differs from that of the lateral components by 90°. This matrix is obtained on the basis of the invariance under the symmetric operations of the M point. The energy dispersions are given by the eigenvalues of the matrix (3):

$$\omega_{+,-}^2(\vec{k}) = \omega^2(0) + \frac{1}{2}\vec{k}^2\Big[(A+B) \pm \sqrt{\{(A-B)^2+C^2\}/2 + \{(A-B)^2-C^2\}\cos\{4\chi(\vec{k})\}/2}\,\Big]. \qquad (4)$$

The angle $\chi(\vec{k})$ is shown in Fig. 1. Here $A > 0$, $B > 0$ and $A + B > |C|$ because the C phase is stable on the clean surface and the two dispersion branches must have the minimum at the M point. The eigenvector on the lower energy branch, $\vec{u}_1(\vec{k})$, has a finite z-component for wave vectors on the Λ line (Fig. 1) according to the EELS analyses of the hydrogen-induced IC phase [4]. Thus $A < B$ holds.

3. Free Energy Functional of the Reconstructed Surface

From the above analysis we can write the general form of the G-L type free energy functional of the reconstructed surface:

$$F(\varphi,\beta) = \sum_{\vec{k}\mu} \gamma\{T - T_{so}(\vec{k},\mu)\}|\varphi(\vec{k},\mu)|^2$$

$$+ \sum_{\vec{k}_1\vec{k}_2\vec{k}_3} \sum_{\mu_1\mu_2\mu_3\mu_4} \delta(\mu_1\mu_2\mu_3\mu_4)\varphi(\vec{k}_1,\mu_1)\varphi(\vec{k}_2,\mu_2)\varphi(\vec{k}_3,\mu_3)\varphi(-\vec{k}_1-\vec{k}_2-\vec{k}_3,\mu_4) \quad (5)$$

where μ or μ_i (=1,2) represents two dispersion branches (μ = 1 for the lower energy branch), $\varphi(\vec{k},\mu)$ the displacive amplitude of the mode $(\vec{k},\mu)$, and γ (>0), $\delta(\mu_1\mu_2\mu_3\mu_4)$ are constants. The $\vec{k}$ dependences of $-\gamma T_{so}(\vec{k},\mu)$ (μ=1 and 2) in (5) are considered to be given by those of the energy dispersions (4). $T_{so}(\vec{k},\mu=1)$ and $T_{so}(\vec{k},\mu=2)$ correspond to $\omega_-^2(\vec{k})$ and $\omega_+^2(\vec{k})$, respectively. $T_{so}(\vec{k},\mu)$ has the degenerate maximum at the $\bar{M}$ point.

4. Effects of Adatoms on the Reconstructed Surface

On exposure to hydrogen each of the half-order LEED spots, which reveals the $(\sqrt{2}\times\sqrt{2})$R45° structure on the clean surface, splits into four in the (10) and (01) directions [5]. It can be attributed to the change of periodicity of surface W atom displacements because adsorbed H atoms contribute little to the LEED intensity. In this section we investigate the effects of the interaction between adatoms and "displaced" surface W atoms including second-layer atoms (A-W interaction) and the adatom-adatom (A-A) interaction on the C-IC transition. The present analysis starts with the adsorbate free energy functional which Lau and Ying proposed [7]. An adatom sees an adiabatic potential which depends on the displacements of its surrounding W atoms. We take account of the A-W interaction only in the first order in W displacements and the A-A interaction up to the next-nearest neighbors (the nearest and next-nearest interaction strengths are denoted by J_1 and J_2, respectively.) for three kinds of adsorption site, a)4-fold hollow, b)on-top, and c)bridge. We use the mean-field approximation and expand $\vec{k}$-dependent terms up to the second order in k_x or k_y in the free energy functional to examine the instability of the commensurate toward the incommensurate structure. Adatoms are considered to be coupled with the displacement modes only on the lower energy branch.

(a)4-fold hollow site

Let us pay attention to the interactions of an adatom with its surrounding "displaced" W atoms. We consider the normal and lateral (diagonal) displacements of the four first-layer W atoms encircling the adatom and the normal one of the second-layer W atom beneath the adatom (Fig.2(a)). We express the total free energy functional, which consists of the substrate and the adsorbate (the A-W and the A-A interactions) parts, in power series of displacement-mode amplitudes. The displacive transition temperature $T_s(\vec{k})$ for mode $\vec{k}$ at coverage Θ (Θ=1 corresponds to saturation.) is obtained by setting the quadratic coefficients to zero:

$$T_s(\vec{k}) = T_{so}(\vec{k},\mu=1) + \frac{a_H\vec{k}^2\sin^2\{\vartheta(\vec{k}) + \chi(\vec{k})\}}{T_s(\vec{k})\left[\frac{1}{\Theta} + \frac{1}{1-\Theta}\right] -4(J_1-J_2) +(J_1-2J_2)\vec{k}^2} \quad (6)$$

where a constant a_H is positive definite and is expressed by the A-W and the

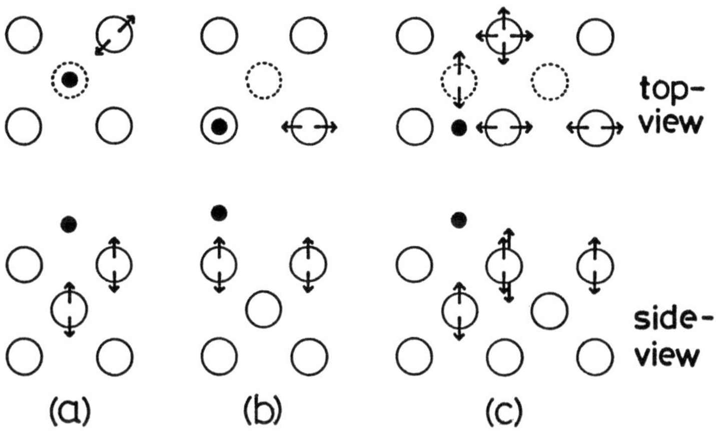

Fig. 2 Interactions between an adatom (filled circle) and its surrounding "displaced" W atoms (open circles). The arrows indicate the W displacements taken into consideration. (a)4-fold hollow site; (b)on-top site; and (c)bridge site

W-W interaction constants, and the angle $\chi(\vec{k})$ is defined by k_x=k $\cos\chi(\vec{k})$ and k_y=k$\sin\chi(\vec{k})$. As the temperature is lowered, the displacive transition takes place at the maximum $T_s(\vec{k})$. The A-W interaction may shift this maximum from the M point, and this effect becomes larger with increase of coverage for $\Theta < 1/2$. If the maximum point deviates from the M point in some direction, the IC phase can become stable at the displacive transition. Nitrogen atoms are considered to be adsorbed in the 4-fold hollow sites [6]. If the observed streaky features "in the (11) directions" [6] are assumed to be explained by the C-IC transition, C < 0 must be satisfied. At least, up to the second order in k_x or k_y, the A-A interaction only makes an isotropic contribution to $T_s(\vec{k})$.

(b)On-top site

We consider the normal displacement of the first-layer W atom beneath an adatom and the normal and lateral ones of the four next-nearest W atoms, as is shown in Fig. 2(b). $T_s(\vec{k})$ is determined by the equation

$$T_s(\vec{k}) = T_{so}(\vec{k},\mu=1) + \frac{a_{OT}\vec{k}^2\cos^2\{\vartheta(\vec{k}) - \chi(\vec{k})\}}{T_s(\vec{k})\left(\frac{1}{\Theta} + \frac{1}{1-\Theta}\right) -4(J_1-J_2) +(J_1-2J_2)\vec{k}^2} \qquad (7)$$

where a constant a_{OT} is positive definite. Under the above-stated conditions for A, B and C, the A-W interaction favors the IC modes on the Λ and Σ lines "equally". The $\vec{k}$ dependence of $T_{so}(\vec{k},\mu=1)$ determines which direction is more favorable. The Λ and Σ directions become more favorable for $B - A > |C|$ and for $B - A < |C|$, respectively.

(c)Bridge site

We take account of the normal and lateral displacements of the first-layer W atoms up to the third-nearest ones and those of the two second-laver atoms nearest to the adatom (Fig. 2(c)). The equation to determine $T_s(\vec{k})$ is given as follows:

$$T_s(\vec{k}) = T_{so}(\vec{k},\mu=1) + \frac{Q(\vec{k})}{P(\vec{k})} - \frac{R(\vec{k})}{P(\vec{k})^2} \qquad (8)$$

where

$$P(\vec{k}) = T\left(\frac{1}{\Theta} + \frac{1}{1-\Theta}\right) -4J_2 + J_2\vec{k}^2 \qquad (9)$$

$$Q(\vec{k}) = a_{B0} + \vec{k}^2\Big(a_{B2} + c_{B2}\cos\{2\vartheta(\vec{k})\}\cos\{2\chi(\vec{k})\} + s_{B2}\sin\{2\vartheta(\vec{k})\}\sin\{2\chi(\vec{k})\}\Big) \qquad (10)$$

and

$$R(\vec{k}) = b_{B2}J_1\vec{k}^2\sin\{2\vartheta(\vec{k})\}\sin\{2\chi(\vec{k})\}. \qquad (11)$$

Here a_{B0} (>0), a_{B2}, c_{B2}, s_{B2} and b_{B2} are constants which depend on the A-W and the W-W interaction constants. Unlike the two preceding cases, the A-W interaction makes a finite contribution to the free energy functional at the M point ($Q(\vec{k} = 0) > 0$). We can not derive any clear result about the C-IC transition because the situation depends on parameter values. According to the EELS analyses [4], hydrogen atoms are adsorbed in bridge sites. For simplicity, here we only take into consideration the lateral displacements of the first-layer W atoms. To explain the splitting of half-order spots "in the (10) and (01) directions", these displacements must be considered up to the third-nearest W atoms. There are two kinds of the next-nearest neighbor pair of adatoms,with and without a first-layer W atom in between. However, even if this distinction is considered, the A-A interactions make an isotropic contribution to determine $T_s(\vec{k})$.

5. Concluding Remarks

The surface displacement modes leading to reconstruction form two dispersion branches degenerate at the M point. The eigenvectors and the energy dispersions of these modes have been obtained. The general form of the Ginzburg-Landau type free energy functional of the clean surface is proposed. The interaction between adatoms and "displaced" surface W atoms has the effect of favoring the IC phase, in competition with the tendency of substrate towards the C phase. This interaction might also play a significant role in C-IC transitions observed in other systems [8,9,10].

Acknowledgments

The authors would like to express their thanks to K. Cho, H. Nagayoshi, K. Makoshi and the other members of their group for comments and discussions. The numerical calculations were performed on the ACOS 1000 system at the Computer Center, Osaka University.

References

1. A. Fasolino, G. Santoro and E. Tosatti: Phys. Rev. Lett. 44, 1684 (1980)
2. T.E. Felter, R.A. Barker and P.J. Estrup: Phys. Rev. Lett. 38, 1138 (1977)
3. M.K. Debe and D.A. King: Surface Sci. 81, 193 (1979)
4. R.F. Willis: Surface Sci. 89, 457 (1979)
5. D.A. King and G. Thomas: Surface Sci. 92, 201 (1980)
6. D.L. Adams and L.H. Germer: Surface Sci. 26, 109 (1971)
7. K.H. Lau and S.C. Ying: Phys. Rev. Lett. 44, 1222 (1980)
8. H.M. Kramer and E. Bauer: Surface Sci. 107, 1 (1981)
9. K.J. Rawlings, G.G. Price and B.J. Hopkins: Surface Sci. 100, 289 (1980)
10. K.J. Rawlings, G.G. Price and B.J. Hopkins: Surface Sci. 95, 245 (1980)

Index of Contributors

Chemistry and Physics of Solid Surfaces V

Editors: **R. Vanselow, R. Howe**

1984. 303 figures. XXI, 554 pages. (Springer Series in Chemical Physics, Volume 35). ISBN 3-540-13315-1

Contents: The Molecular Surface Science of Heterogeneous Catalysis: History and Perspective. - Fourier-Transform Infrared Spectroscopy in Heterogeneous Catalysis. - Magnetic Resonance in Surface Science. - Mössbauer Spectroscopy: Applications to Surface and Catalytic Phenomena. - Heterogeneous Photocatalysis with Semiconductor Particulate Systems. - Laser Studies of Surface Chemical Reactions. - Surface Compositional Changes by Particle Bombardment. - Structure Determination of Small Metal Particles by Electron Microscopy. - Reconstruction of Metal Surfaces. - Surface Crystallography by Means of SEXAFS and NEXAFS. - Determination of Surface Structure Using Atomic Diffraction. - An Atomic View of Crystal Growth. - Ising Model Simulations of Crystal Growth. - Phase Transitions on Surfaces. - Finite Size Effects, Surface Steps, and Phase Transitions. - Recent Developments in the Theory of Epitaxy. - Angle-Resolved Secondary Ion Mass Spectrometry. - Determination by Ion Scattering of Atomic Positions at Surfaces and Interfaces. - Surface Phonon Dispersion. - Intrinsic and Extrinsic Surface Electronic States of Semiconductors. - Work Function and Band Bending at Semiconductor Surfaces. - Subject Index. - Contents of Chemistry and Physics of Solid Surfaces IV (Springer Series in Chemical Physics, Volume 20).

The series **Chemistry and Physics of Solid Surfaces** presents selected review articles predominantly from the area of solid/gas interfaces. These articles are written by internationally recognized experts, the invited speakers of the International Summer Institute in Surface Science (ISISS).
It is the purpose of this series to bring researchers in academia and industry up-to-date and to bridge the gap between conventional surface science textbooks and specialized conference proceedings. Extensive literature references are provided together with a detailed subject index.

Springer-Verlag
Berlin
Heidelberg
New York
Tokyo

MIX
Papier aus verantwortungsvollen Quellen
Paper from responsible sources
FSC® C105338

If you have any concerns about our products,
you can contact us on
ProductSafety@springernature.com

In case Publisher is established outside the EU,
the EU authorized representative is:
Springer Nature Customer Service Center GmbH
Europaplatz 3, 69115 Heidelberg, Germany

Printed by Libri Plureos GmbH
in Hamburg, Germany